THE COMING REVOLUTION

in

PHYSICS

Samuel K. K. Blankson

SAMUEL K. K. BLANKSON

Published by Blankson Enterprises Ltd.,
www.practicalbooks.org. This book is made
available through Print on Demand.

A copy of this title is held by the British Library.

Paperback Edition (9 x 6)
    ISBN 13: 978-1-4452-2809-9

Also available in eBook and hardback format from Lulu.com.

CONTENTS

INTRODUCTION.. 6

THE COMING REVOLUTION IN PHYSICS 17

COMMENTS AND CONCLUSIONS.. 32

APPENDIX I.. 47

    TIME AND QUANTIFIED TIME... 47

    WHAT IS MEASURED BY THE CLOCK? 50

APPENDIX II... 54

    THE PRINCIPLE OF MATHEMATICAL EQUIVALENCE 54

APPENDIX III.. 59

    WHY SPACE ON ITS OWN IS NOT "SPACE-TIME"................................ 59

APPENDIX IV 65

    THE MISCONCEPTIONS OF TIME IN RELATIVITY 65

APPENDIX V 76

    REPLY TO SOME CRITICS ON THE WEB 76

    WARPED SPACE AND CURVED SPACE-TIME....................................... 82

REFERENCES 108

INTRODUCTION

I implore the reader to ponder this Introduction with due care and attention because it has been deliberately designed to convey a clear and cogent gist of my theory of time to such an extent that, if properly considered, even the rest of the pamphlet can be ignored, well perhaps not completely!

Acting on the advice of my agents in Ghana, I published an article in the 31st January 2003 issue of the "Ghanaian Times", entitled The Coming Revolution in Physics as an advertisement for the purpose of promoting two little pamphlets then being published in Ghana. The Newspaper publisher, editors or printer managed to ruin the article with misprints and sheer mathematical ignorance---e.g. printing the square root as an ordinary prime number '2' rather than superscript, to ruin equations, and other errors.

Actually the little books had to be abandoned. They were written by me, and I admit that they were worthless, because, as I later realised, they were based on the Minkowski formula which is logically flawed. In this sense, Einstein bears part of the blame for the present mess in physics. If he had not adopted the Minkowski theory, and even praised it, we wouldn't be in this chaotic situation, where space is supposed to be four dimensional by means of the Minkowski formula--- yet that same formula is described by no less an authority than Professor Sir Arthur Eddington as fictitious and arbitrary. How can we build rational physics, cosmology and astronomy, on a fictitious foundation and expect them to be logically sound as a true reflection of the external world?

I know that scientist will ask about my scientific and academic credentials for daring to question Minkowski and even predict that, because of his formula, physics is heading for another revolution so soon after Einstein (at least the Almighty Royal Society treated me as someone beneath contempt.) If they refused to look at relativity on the grounds that Einstein was a mere clerk, then I suppose when they realise that I am even black as well, they might condemn me to the gallows as a lunatic. The fact, however, is that in science and rational thought logic should rule the human intellect, and logically the Minkowski formula is false, not only defective, but completely false. Let me explain that by 'logic' or 'logical process of thought' I am referring to the process whereby we learn of the nature of the external world. We do this in strict accordance with evidence to the senses. How otherwise could we live in consonance with nature? You can drop into a river, but you cannot drop onto a large rock or cemented floor looking like a river; the difference between the two must be known beforehand. Anything less than the facts of nature is fantasy and properly ignored except by the gullible. Anything more, looking ahead into nature, sometimes far ahead, is called imagination; but even that, to be useful, must accord with known facts about physical reality.[1] The whole of advanced mathematics and the higher levels of logic are justified on this grounds. Thus Minkowski sought to make his theory acceptable by complex mathematics; yet however complex, his ideas are based on imaginary time coordinates. In the study of

_____

[1] Hence Bertrand Russell and Professor A.N Whitehead suggested that our knowledge of the external world comes to us by means of 'a construction', or logical construction, not logical inference alone. For the benefit of the non-professional, let me explain that this is a new philosophy of physical reality, carefully constructed to accord with the facts of physics; and it is done so successfully that, in my opinion, the photon, or the particle of light, becomes 'What Is', due to the quantum or QED.

time alone, I submit that this method is unacceptable because time is not imaginary. You must quote time before you can use it to reason; and the only way to quote time is by using the clock, because, as argued below, the metaphysical nature of time can never be known; the nearest we come to that knowledge is how it passes---second, second, second, or year after year after year. We know of time in units only, as established with repetitive cycles. So creating the unit or units of time, the year for instance, is creating our time in essence.

Now, a four-dimensional geometry incorporating time is not feasible because the time comes to exist only as a product of points as applied to space (second, second, second relying on points); it cannot be naturally part of space because we use the space to create it in the first place otherwise it wouldn't exist. It is produced by space, and therefore it can only be an offspring of space. It is a third entity to points and space; otherwise we couldn't have it in units. The earth-year, for example, is produced by points and space: from the Meridian point the earth has to travel through space back to the Meridian point before a year can materialise. So the year, as a unit of time, is produced by the space around the sun and the Meridian point; that is why we divide the year exactly into seconds so that about 31, 536,000 seconds equal one year to coincide with the completed orbit of the sun—and start again.

The Minkowski formula can be refuted in a rather mundane or even plebeian manner. There is no need for complex mathematics to confuse the issue. Time is used by everybody; as far as possible it should ne discussed in a manner that almost everybody can confirm some of the facts through his own experience. Earth time is universally accepted, so we can all agree on that as the basis for reasoning about time; yet the year is discrete time proceeding unit by unit. Accepting the year as a valid unit of time implies acceptance of the existence of discrete time, or of the fact that earth time, since it is based on the year, is discrete. So, then, the year is one unit of time, replicating itself repeatedly to reach the centuries. It means time does not march; only one year is repeated over and over again. One unit of time increasing by replication is discrete time. Similarly, the subdivisions of the year, all the way to the seconds, are also obtained from the year with points, and are therefore nothing but discrete time. Discrete time does not exist beyond its unit---that is, beyond a single unit of it.[2] The year, for instance, ends on 31st December. When the unit of a discrete time is expended, the time is no more. Therefore discrete time cannot be in every space so that s=ct as given by the Minkowski formula.

Let me point out, too, that accepting the year as a valid piece of time means we know of time only through how it passes and not what time is, since the year gives no idea of the metaphysical nature of time---nobody knows that---all we know is that the passage of the year is the passage of time.

---

[2] The great unsung hero of 20th century physics and philosophy, Professor A.N. Whitehead, made this point abundantly clear when he said, "...a moment of time is to be identified with an instantaneous spread of the apparent world...A time system is a sequence of non-interacting moments [however that moment is defined, long or short---like the one year for instance, subdivided down to the seconds that constitute our time system] *The Principle of Relativity*, Ch. IV. But at the time nobody took any notice as Minkowski was all the rage as the greatest thinker who made relativity accessible to mathematicians. Any mathematician who can only understand relativity through the Minkowski theory is lost.

There is no conceivable other way in which we could produce time in units. Otherwise time is not, and would not be, continuous. Yet all other units of time used on earth are derived from the year with points and space, including the cesium units, since they have always to be related to the second to make temporal sense. In other words, we need regular, or repetitive, cycles to produce time units, and we are using the earth's regular orbits of the sun for that purpose. The metaphysical implication is obvious: time is known and used only in specific units that must rely on points for their creation. See APPENDIX I below---time is quantified time or the word 'time' is meaningless; you can't do anything with it as 'time' unless it is stated in numerical units.

I have made plain in the book that we can never define time; metaphysically nobody can tell what time is. We cannot even define the year logically; other units derived from the year can be defined in terms of the year; but the year on its own is not definable. The best we can do is to show how time passes by means of cyclical or repetitive motions, the year for instance, subdivided down to the seconds. But knowing this gives us a hunch regarding the nature of time. It appears to be the product of motion and physio/chemistry and a means whereby man is joined to nature. You have to be in existence and want to relate to nature in harmony. So you use cyclical motions as your rates of time: so many cycles means so-and-so. But the year is so long that we have learnt to subdivide it down to the seconds and the cesium units and so forth. However, sentience, a theory of numbers and the ability to count are absolutely essential.

But meanwhile there is another theory of time which can also be called 'space-time'. Hence, after all, we have something to thank Minkowski for, since he was the first to use the term. But the new theory of time is owed to

Albert Einstein's incomparable genius. It is the Lorentz local time notion which he extended to all nature. It is the most revolutionary idea of all time---in all things and in all life, even in the universe, et al. Einstein did not set out originally to redefine time; he was not even researching time. And let me add that the units of time are always the same (whether it is local time, natural time, religious time, or whatever); even the Minkowski space-time is also the same, for the second is always the same, *in as much as we continue to use the earth-year as our primary unit of time out of which all other units are derived*.[3] Thus the Minkowski formula works (even though it is arbitrary) because his units of time are still the same, being related to the earth-year. The problem is that, even before the Communist takeover, this crafty Russian sought to dominate the world with a system of time cunningly based on the earth-year but only supported, logically, with arbitrary elements---which are unacceptable.

So it is not necessary to credit Einstein with creating any new sort of time. The time units are still the same. He did not deliberately consider time at all. His attention was attracted by the Lorentz discovery of t^1 or local time. For it was an extraordinary idea, quite impossible on Newtonian theory of time. Here was a different sort of time. The first occasion that time has been seen to be different in different situations in science, whereas it was supposed to be general and absolute.

---

[3] The situation would be different if we stop using the earth-year as our primary unit of time. For this reason, I suspect the quantum would be different in other parts of the universe, because it materialises on earth by means of our unique periodicities since it is time-dependent; and these periodicities are bound to differ from place to place.

But then Einstein made matters worse by describing this strange and shocking new time of Lorentz as, 'time, pure and simple'. The implication being that it was the true nature of time; that every time is somebody's local time and that time is neither absolute nor general, but rather dynamic. An intellectual, scientific, philosophical and religious bombshell of incalculable consequences. There is no greater theory in science, except perhaps Darwin's evolution concept simply because it concerns the origin of life itself. We are still pondering that strange notion of time.

For a start, it means time can begin from many places, and Einstein even went on to suggest that 'there are as many times as there are bodies in the universe'. Thus, logically, s cannot be equal to ct. Space cannot already be time. If every body in the universe has to have its own time, and since every body is in space, then space (general space in the universe) cannot already be time.

Why does every body has to have its own time if time was already in the very space in which it is existing as s=ct? This is an example of the simple ways in which logic can be used to expose errors in mathematical thought, and I mean the sort of complex mathematics employed by Minkowski: for the plain truth is that any body will have to create its time first, since the creation of time involves points, and the point (and the intellectual use of points) is well-known to be a human invention? It also means that, by implication, space cannot be time without a conscious effort to create the time for that particular body concerned. And what applies to one body applies, naturally, to every other body in the universe---without racial bias, or (religious) favouritism!

Unfortunately, Minkowski did not specify that his s=ct equation applies only to the earth's space so that, in his defence, his supporters could argue that he was only referring to the earth, even if such a defence could be logically sustained. All we know is that, due to his concept of 4-D space, Mathematicians and cosmologists always refer to space (every space and all space), as 'space-time', while all time, too, is 'space-time'. This makes it easy for them to propound theories on the basis that time and space are unified. The truth is that they carry earth time's periodicities in their heads to apply to any space. This is fraudulent, or in polite scientific language, according to Russell and Eddington, only arbitrary, because, according to relativity, earth time is not applicable to other parts of the universe---in the words of Russell, 'without ambiguities'. Even worse, the equation upon which s=ct rests is itself relying on imaginary time coordinates.

Often we are told by some science writers that Einstein was researching time because its nature was creating difficulties in his train of thought. That is absolutely untrue. We need proper historical analysis here. He wanted time in his 3+1 formula, but what time, whose time, if time 'is' not absolute and general? He was shocked by the Lorentz discovery; and then his genius made all the difference---all time must be somebody's local time; therefore time can begin from anywhere.

How does anybody's time begin? Obviously it has to begin with the application of points to space otherwise we couldn't have it in units.

As the title of this little books implies, as a result of all this (specifically due to the nature of time), I think physics is heading for total collapse in the very near future, because over the years since relativity, all theories have relied on the concept of 4-D geometry---but it simple does not exist as part of the natural apparatus of the universe; it came to exist solely by the mathematics of Minkowski. I object that mathematics can be used to alter physical reality. See APPENDIX II below. Fortunately, in theoretical physics when things go wrong nobody gets hurt; and so we shall survive.

The irony is that the scientific establishment is fully aware that the Minkowski theory is false: (1) Because, obviously, i cannot be used to invoke or represent time as a whole in the sense as Einstein himself has stated below. I had to be careful with this. I realised that unless the name of Einstein is associated with it, nobody would look at it. And he states clearly, that, under the Minkowski system we start by representing time with i. (2) In his mathematical Theory of Relativity, (page 9), Professor Sir Arthur Eddington called the Minkowski formula fictitious, but added that "we shall continue to employ it...". I have shown why scientist want to continue to employ it; but the critical point is that it vitiates their theories, that is the problem---logically most of physics is thereby distorted so much that philosophers have difficulty interpreting physics in epistemology and ontology. Scientist just state s=ct and move on; but s is not equal to ct, because the Minkowski ict equation purporting to equate space to time is logically untenable. It is typical of formulaic mathematics that, so long as s=ct is in the textbooks, scientists will simply cite it and build their theories upon that, as if they don't know how to think. By implication time is supposed to be one and the same thing as space, not as is seen in nature, but by the mathematics of Herman Minkowski---that is the irony of the situation. If you object, they accuse you of insanity. Or, like the Royal Society they will treat you as if you do not deserve a common acknowledgement of your submission. A venerable institution like the Royal Society should learn to respect contributors whether they are black or white. (3) Above all, the great Bertrand Russell also called the Minkowski theory that is now dominating physics arbitrary.

Well, I am nearly 72, and according to Einstein himself, at this age, beyond 26, nobody is supposed to be capable of much basic work. And The Royal Society may think I am beneath contempt, yet I am still there, churning out my defective little books against the Minkowski fiction that, I think, is distorting relativity and physics as a whole. I maintain that the Lorentz local time concept adapted by Einstein, when he said it may be defined as 'time, pure and simple'[4] is (or was) obtained with the application of points as applied to space; so we can call it 'space-time'. It makes time necessarily discrete--- being the product of points, otherwise we couldn't have it in units: the year for example. When it comes to the sub-division of the year all the way to the seconds, the situation is clarified, and we realise that the sub-divisional units can only be discrete.

The concept of discrete time has momentous implications. It means time does not march; history is not the march of time but of events. Discrete time cannot curve; so there is no such thing as 'curved space-time' by which many scientific writers claim that time travel is 'a scientific possibility'. Due to Einstein's discovery of the true cause of gravity by means of curved space, the Minkowski s=ct equation has distorted all physics since then, by misleading scientists into thinking that when space curves it takes time with it. This has spawned fantastic suppositions in general relativity so much that cosmology is in a mess; the universe is interpreted in so many imaginary ways that even god will be confused

---

[4] I am surprised that no writer has noticed the importance of this remark by Einstein. In my estimation Einstein was so great, so good and so clever that, to me, everything he said even in jest was important. His description of the Lorentz local time means all time is somebody's local time: time can begin from anywhere, and by anybody, or any 'body'.

by his own 'Creation'---exactly as Einstein put it, namely, that since the mathematicians invaded his theory he could not even understand it any more.

Furthermore, discrete time can only pass by and seem continuous through the procession of its units---second, second, second; or year after year after year, and similarly for all the other units of time.

Finally, the reader will notice that, apart from this Introduction and the Appendices culled from published sources, I have done away with complex academic footnotes, as I want him or her to find this an easy read like an article in a newspaper. I conclude the book with five Appendices that expand some of the difficult points in the book. They have been used in my other books; but I offer no apology for including them here to make my argument easy to understand. Time is not only difficult but an impossible subject; those of us writing about it deserve sympathy! We still cannot tell the metaphysical essence of time; but at least we now know that it began on this earth, and passes by and seems continuous through the procession of its units---second, second, second, or year after year after year.

SAMUEL K.K. BLANKSON

ACCRA and LONDON

THE COMING REVOLUTION IN PHYSICS

PART ONE

THE PROPOSITIONS

1. Mathematics cannot alter natural physical dimensions; it can only mirror what is already there.

I believe relativity has been contorted and therefore undervalued and made to look vulnerable for easy criticism in philosophy and theology by the fact that everybody refers to space as 'space-time' and time too as 'space-time' in the mistaken assumption that space has been successfully equated to time with the Minkowski 4-D geometry. It is comically ironic that scientists, as intelligent as they are, can believe that mathematics can be used to alter physical reality from its obvious 'real' nature of having three dimensions plus time in the 3+1 continuum of Einstein's original relativity conception of physical reality, to one of four dimensions where time is incorporated into space successfully! Yet that is what physical reality has been converted to in the name of Einstein and relativity.

2. Einstein was probably coerced to accept the Minkowski formula for equating space to time. Current theories about muons cannot be true because they are based on the Minkowski equation of space to time which is logically flawed.

In fact, originally Einstein called the Minkowski formula 'superfluous learnedness', and moaned that because of the mathematicians he could not even understand his own theory anymore. Even worse, these same 'objective' scientists and mathematicians refer constantly to "The beginning of time", and "The dawn of time", in total ignorance of the fact that space-time is discrete, and discrete time can have no beginning or dawn stretching from its own metaphysical genesis.

Perhaps they are still thinking of time as part of the divine 'creation'. Or perhaps what they mean is "The beginning or dawn of existence". There can be no 'dawn of time' because technically time does not exist in nature naturally without sentience to count the orbits of the sun as years, for instance. For if we do not count the orbits as years, there are no years. What we call 'time' is a mathematical contraption that merely helps us to know how time is passing, not what it is. Nobody knows what time is. Yet the idea that time had a 'dawn' has sunk so deep into culture that, added to the so-called curved space-time and Time Dilation theories, the inferences that can be derived from this formula of reality is literally limitless. They are all based on the Minkowski equation of space to time---which is false.

The situation is sad, literally sad and unfortunate, for sometimes it makes some scientific passages read like magical texts---especially those concerned with the mystery of how we find muons on earth. Let me tell it as a story, I am sure the reader will find it interesting.

These tiny, ephemeral nuclear particles should not last long enough to reach the earth, but they do. Mathematicians explain this with the theory that the Minkowski formula means every 'thing' which occupies space has time in its basic nature. This time slows down with speed---therefore the muons can last long enough to reach the earth from the distant stars whence they come. Excellent. But if the Minkowski formula is arbitrary, then this explanation is false. Yet it is now the received wisdom about muons in mathematical physics---are we serious?

3. *Why gravity cannot influence time, if space is not equated to time. Also, if that is not the case, then*

time travel is not feasible, and curved space-time does not exist. These Minkowski ideas make physics and cosmology rather simple. Unfortunately nature is not that obliging. Its mysteries will forever remain mysteries to the end of the world, (and we know that the world is going to end 'soon' by geological time), because human beings are not important creations vested with powers of investigation. We are nothing in the vast cosmos; our little achievements are but ripples at the shores of the vast and incomprehensible universe of no conceivable limits.

In this little pamphlet I am using only two inferences to illustrate what I regard as 'the basic error' of the Minkowski 4-D geometry at the heart of mathematical physics, knowing that it is actually regarded as the best interpretation of relativity and therefore I will probably be accused of insanity, even though Professor Sir Arthur Eddington and Bertrand Russell condemned it as arbitrary. First, I will discuss relativity and gravity from a book review in The Sunday Times, a major Newspaper, thus popularising the mistaken notion more widely. My second example of how relativity is used mistakenly to confuse us all about the nature of physical reality as conceived by Einstein is about time travel, taken form a successful book originally published in Australia. When a book crosses Continents it means some leaders in science believe in its importance. In any case, I can only comment on these two topics; the serious matter of space-time requires a genius with infinite mental prowess to expunge its corrupt influences from our intellectual life. It is sad, really sad, I repeat, and damaging too.

In James McConnachie's review of In Search of Time (The Sunday Times, 'Culture', p.48, 19th July 2009), he states: "Because gravity slows time, an atomic clock in Boulder, Colorado, at 1,600m above sea level runs a whole five microseconds faster…"

I am afraid, theoretically all such experiments are fundamentally flawed because gravity can 'do' no such thing in nature, since it is not conceived as some sort of a meddling sentient being with a mind of its own and infinite tentacles. Some people want to march the rest of us blindfold through mythological realms of their imagination and call it 'science'. As a result of the efforts of the mathematical interpreters of relativity it now requires great intellectual efforts to sort out what Einstein said and what he could not possibly have said. It does not matter how they achieve what they claim to have achieved---there seems always an element of religion in it. For the simple fact is that theoretically, or logically (that is, in theory and in logic), the results claimed contradict relativity. There is no logical mechanism through which gravity, conceived as a mere physical effect in nature caused by the curvature of space, can affect the hands of clocks, that is, 'all clocks', so as to alter time per se on this planet or anywhere else; also, there is no temporal control of physiological processes by means of clocks; time will make you age but not by the performance of any particular clock. Obviously belief in pre-relativity time dies hard because it is mentally connected to the religious instinct. Otherwise what kind of time can be so affected universally---how do they define it without the universality of time abolished by relativity? They do not talk of events between two clocks which may be considered as 'possible'; they claim that it affects all time, or time per se. In this particular instance, the claim is that a particular clock runs faster---what do they mean by 'faster'? Faster as against time on earth as a whole? It is just not possible. There is no logical mechanism for it. And since there is no standard time frame, there is also no 'fixed' time with which it can be compared to determine whether it is faster or slower. If, on the other hand, it is a mere comparison between two clocks, then it must on no account be regarded as part of natural physical law, or in anyway capable of its alteration---that is not how we change the laws of physics.

The people who perform these experiments (together with all others about time travel 'curved space-time' and the claim that speed slows time), assume on the Minkowski formula, even before they start, that space and time have been successfully unified with his mathematical proposals. In fact, they are not because time requires points---points in application to space for the creation of quantified time, which thus makes time the product of points and space, and, as such, independent of space and so can never be linked back to space again. Without 'quantified time', which is explained below, we could never tell 'how much time?'. So since the uses of time centre on 'how much time?' in numerical units, scientific time must recognise only quantified time; otherwise time can be defined a thousand different (mystical) ways.

4. How the concepts of clock paradox and time dilation are misused in physics and false definitions of time in science generally.

In case the reader is wondering about my answer to the 'science' behind all this, here is my answer. The so-called 'clock paradox', the Minkowski formula for equating time to space and the concept of time dilation, constitute the basic grounds upon which the interpretation of time as going slower when speeded up is based. The funny thing is that nobody can make time go slowly; what is found is that under certain conditions clocks run 'apparently' slower than other clocks. You cannot, and must not, build a metaphysical or religious interpretation for the whole of time on this, for normally one scientific fact does not alter the philosophy of physical reality. But never mind. Let us look at these events in some detail. First, the clock paradox. This is not part of relativity because Einstein himself said so. It is true that it has been observed, but he said it cannot be used to contradict relativity. Here I quote from Abraham Pais's famous biography of Einstein and his science ('Subtle is the Lord...', Oxford, 1982, Ch.7.): "Einstein rather casually [my italics, meaning 'not very seriously'] mentioned that if two synchronised clocks C1 and C2 are at the same initial position and if C2 leaves A and moves along a closed orbit, then upon return to A, C2 will run slow relative to C1, as often observed since in the laboratory. He called this result a theorem [my italics, meaning 'a curious fact yet to be explained'] and cannot be held responsible for the misnomer clock paradox, which is of later vintage. Indeed, as Einstein himself noted later [E6--- this refers to the basic paper in which Einstein wrote this]. 'no contradiction in the foundation of the theory can be constructed from this result' since C2 but not C1 has experienced acceleration." End of the mystery; end, too, of the religious view that all time slows down upon the performance of one clock which experiences acceleration. Next, time dilation. It is often said that this was predicted by Einstein. In fact, it was a mystery discovered by Lorentz which Einstein helped to explain with his theory of frames. I personally suspect it inspired the theory of frames in essence. The dilated clock has no control of all time; it has control only of its own performance, and strangely enough, its performance in its own frame remains normal since those carrying it will notice no change in its performance; but those looking at it in the moving vehicle from outside will notice that it is running slowly. So the anomaly is called 'the dilation of time as a measure of moving clocks'. Again, the mystery is resolved. Now the Minkowski formula for equating time to space. Professor Sir Arthur Eddington called it arbitrary and fictitious. It is not science at all except that, as explained below, mathematicians prefer it to the difficult and somehow philosophical 3+1 formula proposed by Einstein.

To recap: The hand of a speeding clock will run slowly (due to the acceleration and not because of any other known or unknown mysterious influence). Thus there is no physiological connection between man and any clock; the slowed clock cannot affect the ageing process of anybody. Yet still some people jump from this to the conclusion that at high speeds man's ageing process will also slow down because man is in space and, according to Minkowski, space and time are unified into one entity. But the Minkowski formula that equates time to space has been described by one of the pillars of science as 'arbitrary and fictitious', as quoted below from Professor Sir Arthur Eddington---Bertrand Russell also called it arbitrary. So neither the clock paradox nor the Minkowski formula has any relevance in serious discussions of time such as we are engaged in at present.

Time dilation is completely irrelevant. The current received wisdom of those who are not completely conversant with relativity to the effect that time dilation is "The principle that intervals of time are not absolute but are related to the motion of the observer"[5], is complete nonsense. It is one of those annoying but popular misconceptions in physics that inspired me to write this pamphlet. As I have said, sometimes they make passages in physics read like magical texts. Some writers have imagined their own flawed interpretations of relativity and are peddling them as 'science'. The fact that intervals of time are not absolute is a concept in no way related to time dilation. That was a discovery of Einstein by means of a completely different line of thought or logic. It was Lorentz who discovered time dilation, calling it 'local time'. He could not think of

------

[5] Oxford Encyclopedia, OUP, 1998, p1336

what to do with it, and literally put it as side; as far as he was concerned, it was not 'true time'---see below. He was right; it is not crucial in physics at all. Its importance is philosophical or inspiration; it led Einstein to his greatest theory---that time is not absolute but dynamic and differs from place to place. It made him realise that time can begin from anywhere and by anybody for any purposes, and that time is neither divine or cosmic in origin. As such it cannot be the same as "existence"; it is additional to it. Rather it is something that accompanies or complements existence, since sentience is required for classifying the orbits of the sun as 'years'. Because of its nature we have to be very careful with time; it seems cosmic or divine; yet it is man who calls the orbit of the sun as 'a year'---his basic unit of time. The religions are very crafty about things. You see, to overcome all this, they claim that it is god who gives man all the powers he seems to have, including the ability to invent his time! In truth, Einstein's analysis of order and simultaneity were the methods that led him to his theory of time, which I consider his greatest achievement, due to the mysterious nature and supreme importance of time. Bertrand Russell too considered it as 'perhaps' his most important discovery. Personally, I have no doubt about it.

5. The true scientific facts about gravity and time; and why gravity can never influence time---except in theories where time is (only) falsely equated to space.

The true scientific position with regard to time and gravity in relativity (by which we get our most rational notion of physical reality), is precisely as stated by Professor Bernstein in his book Albert Einstein and The Frontiers of Physics (Oxford, 1996, p.110): "In the absence of gravity, space and time are distinct entities. In the metric of special relativity, they play distinctive roles [so much for those who repeat the myth that special relativity included the Minkowski formula for space-time]... But in the presence of gravity the metric is altered, and space and time become mixed up with one another...The metric has four coordinates, but the space and time coordinates become entangled..." This refers to the strong gravitational field, which is not encountered on earth; so it is not even relevant here. One popular Reference Book on science states clearly that special relativity is called "special" because "it deals with conditions in which gravitational forces are not present". To emphasise this, Professor Bernstein adds that: "Only when gravity is weak can they [space and time] be distinguished in a useful way".

That is the situation on earth as I write, so that we have our stable time; but those who accept that time and space constitute one entity, as Herman Minkowski proposed, believe that time can be contorted by gravity and space. As a matter of fact, even this statement from Professor Bernstein is not strictly correct, because if the Minkowski equation of time to space is false (or in polite scientific language 'arbitrary') then even the strong gravitational force cannot distort time because time remains independent of space as we have always known it to be: time is time in the clock; and the clock is independent of space. Time in the air, as 's=ct' is Minkowski's imaginary time which should have no place in science.

6. Why all the problems in physics since Einstein (or most of them), are caused by the Minkowski false equation of space to time by means of his imaginary time coordinates.

Those thinkers who accept the Minkowski theory are creating a body of physics that is not in congenial accord with physical reality, and their concept of time is misconceived. And if that is so, then several theories in physics and astronomy are inherently flawed. One of these may be the attempt to link general relativity to the quantum, since the quantum is time-dependent---but by whose time? I suspect that, since the quantum materialises by our unique time, it may not be a universal unit of energy throughout the universe but something we get only by our unique time; yet gravity caused by the curvature of space is apparently universal, a universal occurrence. Time is crucial to all phenomena. It is wrong to use earth time in the metric of general relativity because it is a different frame: the physical postulates that make earth time what it is are not present in the metric of general relativity. Those who do not as yet believe that Einstein caught the essence of physical reality must think again. Another of the problematic notions that involve time is the concept of singularity. Since we cannot use earth time in the core of a black hole, what time is there to fuse with space—i.e. from which body's physical postulates is the time obtained since every body in the universe has to have its own time? Evidently, in general relativity there is no time since it is 'a highly unstable metric' in which nothing is certain except the curvature of space; using earth-time's periodicities carried in the head to apply to the metric of general relativity (or any other body in the universe, as Russell has averred---see below), is logically untenable because it violates the Einstein theory of frames.

7. **All books claiming that time travel is 'a scientific possibility' are successful because they assert that Einstein predicted it. In fact, they are wrong; but their books sell millions because the gene in man likes to hear that religious idea. Time travel was never predicted by Einstein, and it certainly is not a scientific possibility. It is bogus science based on the Minkowski fiction that time can curve together with space to no end. From this they infer that it could curve infinitely backwards, or forwards, and add the**

rest as it suits their fancy or religion. Almost all human woes are religiously caused.

Briefly (and I mean very briefly indeed, as I cannot go too deeply into the mysticism spawned by the Minkowski arbitrary equation of space to time), in her book Einstein's Heroes, Robyn Arianrhod wrote: "...suppose you leave the Earth in a very fast space ship, journey through space for seven years, and then head back to Earth; the equations of special relativity predict that on returning home, you will find that everyone will have aged not fourteen years, as you will have done, but fifty. According to your Earth-bound friends, time really will have slowed down for you..."

I want the present reader to ignore all that nonsense about "the equations of special relativity". No such equations are to be found in Einstein's theory; what he said about the so-called clock paradox, as explained above, does not amount to 'equations about time travel'. As Professor Bernstein has stated above, in special relativity time is not equated to space---they are distinct entities: space is space and time is time, with the sole proviso that the time is dynamic, not absolute, and changeable from place to place, or frame to frame, exactly as Lorentz had discovered, and which he called 'local time'. Einstein then extended this notion to the total absence of a universal time. Then the problems began. For, if so, then how did our own time begin; how are we to define it metaphysically (so as to fit it into the scheme of nature, or of physical reality, our theory of knowledge of the external world), and whether or not it can be applied to all nature, and so forth.

The best answer to the writers of this kind of scientific mysticism from which they infer the existence of time travel, is what I have christened 'The Max Planck dictum' (or logic) about time travel. It goes like this: while he could not fault the mathematics of those very clever people who claim that time travel is possible, he pointed out that the basic premises upon which they build their theories are incapable of proof and therefore their time travel suppositions are unworthy of consideration as science. Similarly, the Minkowski formula for equating space to time is based on arbitrary assumptions, being his imaginary time coordinates, since there is no such thing as imaginary time; therefore his theory is unscientific, and all the experiments and inferences based on it are sheer humbug.

8. How the 'science' of Minkowski and his followers has led to a complete misconception of what relativity means.

That clever, dogged champion of science, the famous British writer John Gribbin, tells us that, "Minkowski's geometrical description undoubtedly improved the clarity of the special theory and it is still regarded as the best way to understand it." (New Scientist, 2nd Jan. 1993, p28.) Of course, he is right. There is a problem with relativity in this aspect of the theory---i.e. with regard to the definition of physical reality, but how the mathematicians put it is completely unacceptable because the Minkowski solution which they adore so much only substitutes one kind of abstruse concept for another, but this time with his (Minkowski's) logically untenable imaginary time coordinates.

It is all rather elementary. Man cannot fool nature by dreaming up a quantity by means of a mathematical device known as i, and say that is what reality is. If it is not, the whole system is falsified; and that is what, I think, is happening to physics at present. Accusing critics of lunacy helps nobody, especially the dreamers, the physics, astronomy, and the cosmology built upon the false premise.

Obviously i is quite useful in mathematics, but not in the study of time because time is not completely mathematical; we use mathematics merely to quantify time; otherwise time is going silently even as we sleep. But on waking up, to tell 'how much time' we need mathematics, that is all the connection between time and mathematics[6]: mathematics is used merely to quantify time to make it universally applicable on this planet, as explained below. Time is also cultural. In other words, to be able to tell the time at a glance there must be the 'culturally-created clock'. Yet even Einstein was persuaded to accept i of the Minkowski ict equation in place of the whole of t---mental, cultural and mathematical. See Part I, section 17, of his seminal book 'RELATIVITY' (Reissued in the Routledge Classics, London, 1993.) He wrote: "...we must replace the usual time coordinate t by an imaginary magnitude..." No, Albert, for God's sake, no! Logically this is untenable; that is why Minkowski is bad for relativity and physics as a whole; and only God knows how long it will take physicists to realise this. Creativity in physics, cosmology and astronomy is being destroyed by the slavish and intolerant attachment to the Minkowski theory for equating space to time. The intolerance and scorn poured on critics of the formula ensures that only space-time, as 4-D geometry is tolerated; everybody else is accused of ignorance of counterintuitive mathematics.

I agree that the complexity of the Minkowski mathematics is impressive; but this is the logical basis of the whole structure—and it is logically wrong. Precisely

---

[6] It is very important, though, for all we can ever know of time is how it passes, or how many units of it is going, have gone, or will go---the years, for example; and the mathematically induced units are what we count to know how time is passing---second, second, second, and so forth.

as Einstein puts it, the whole Minkowski formula is based on imaginary time. We start by replacing time, as important as it is, with imaginary time. Is that what we want for the foundation of physical reality in physics? Is it right? Yet this is all the technical grounds advanced to justify the Minkowski formula. Thereafter s=ct—all space is automatically time. When any theory is called 'relativistic' it means it incorporates s=ct as part of it. We are fooling ourselves, not nature; nature cannot be fooled; it will show, eventually, that our physical theory based on the Minkowski formula has falsified subsequent theories.

The philosopher needs not say more than this. As Planck would have put it, philosophers might not be able to fault the mathematics, but its foundation is illogical, therefore the whole structure is indefensible and has no place in physics. Time is not imaginary. Good Lord, unavoidably it's making me, you and everybody age every second! And how can we construct physics on the basis of imaginary time, especially when we are asked to accept that by this imaginary time space and time are fused into one entity as the basis of physical reality? Physicists must realise that if this is not correct, then they are distorting physics in homage to Minkowski, not Einstein. The argument that these distortions occur from time to time is no defence because the fault in the Minkowski formula is glaring.

The interesting and profound query regarding relativity and the definition of physical reality, which the mathematicians claim to understand only with the aid of the Minkowski formula, is this: in the 3+1 formula man is the one to add the time, so that the observer becomes part of the observed. But Einstein has shown that time is not universal; so it means we will have to invent our time and then add it to space or phenomena. This makes relativity somehow philosophical. It would be much simpler, and more scientific, if the time was incorporated into space naturally. In other words, some thinkers, writers and mathematicians don't think the idea of creating or inventing our time to add to phenomena is completely objective or scientific; they prefer the Minkowski incorporation of time into the fabric of space---even though with his imaginary time coordinates which render his formula arbitrary. The other objection would seem to be religious. It seems people are afraid that if there is no natural time then time was not created by God---and therefore the last hiding place of God after Darwin will be gone and no longer available to theologians. A clear indication of the intellectual merits of theology.

PART TWO:

COMMENTS AND CONCLUSIONS

1. The importance of time in life. The true nature of it as far as we can discover. The role of Einstein in all rational discussions of time; and the judgement of Professor Sir Arthur Eddington with regard to the logical merits of the Minkowski theory.

Life consists entirely of two things: Being. Existence (or life), and the necessary time for the regulation of the existence. This makes time the second most important consideration on earth, second only to the life itself. Until Einstein we took time as something of a divine bounty, which was eternal (always there), universal, and absolute (always and everywhere the same); and therefore literally the last hiding place for God, following the Refutation of Idealism and the rejection of 'Creation' by Darwin. You could not deny the existence of God unless you could explain how time came to be.

2. Albert Einstein changed all that. In the very strong words of Professor Sir Arthur Eddington, "Prior to Einstein's researches no doubt was entertained that there existed a 'true even-flowing time' which was unique and universal...Those who still insist on the existence of a unique 'true time' generally rely on the possibility that the resources of experiment are not yet exhausted and that some day a discriminating test may be found. But the off-chance that a future generation may discover a significance in our utterances is scarcely an excuse for making meaningless noises." (The Mathematical Theory of Relativity, Ch.1.)

Einstein showed that time is rather dynamic, and derives from space; it is therefore 'local' in nature, and not at all absolute, but variable from place to place. Since it derives from space points are implied, otherwise we could not have it in units; yet time is known and used only in units. This makes time 'necessarily discrete'. Einstein was right because anything created with points must be a product of space (hence 'space-time') and necessarily discrete. And that is what we find. The basic unit of time is the year from which all other units are derived, but it is one unit of time; and it relies on points, i.e. the Meridian point or line. The yearly cycle is a human concept, relating specific units of space to the sense of duration in the mind (thus we know in the mind that a second is shorter than a minute, etc.) The Meridian line is the beginning and end of our basic unit of time, the year---but there it ends, together with all of its subdivisions. If the earth does not go on, how could we obtain another second in fact---that is, in physical reality or in consonance with reality? For the rational or scientific study of time shows it as something we invent to accord with the features of physical reality, that is why time is important: 'day' means so many hours of daylight; 'night' means so many hours of darkness, with clear cultural implications as to what can or cannot be done during those periods; similarly an hour, a minute, or a second indicates 'a specified distance' from day or night, and so forth, all the way to a year.

To have more years (and therefore more time from nature) we orbit the sun again and again all the way to the centuries. But, as stated above, it must be emphasised that the orbit of the sun is a physical activity; it is man who uses it as a time unit by linking it to the internal sense of duration; so that every unit of space is given a period in sense, or in the mind. We do this so as to create quantified time as opposed to the silent passage of existence, which is not usable as time until it has been quantified into numerical units. As far as I am concerned, only quantified time is 'time', because that is what is available objectively for all and sundry to use. Furthermore, the postulates that give us our time are limited to the earth; so earth time is applicable only to the earth, since time can begin from anywhere and different in different places.

3.   Relativity is everywhere distorted and misunderstood because of the Minkowski formula, but let me explain very clearly that the whole new idea of time came from the Lorentz discovery of t1, a different kind of time, which puzzled him because time was supposed to be universally absolute---that is, general and the same everywhere. He was so baffled that he called it 'Local Time', not the true time, and put it aside. In the words of Abraham Pais, "He proposed to call t the general time and t1 the local time [L16]. Although he did not say so explicitly, it is evident that to him there was, so to speak, only one true time." In fact, Lorentz had discovered that time can begin from anywhere and that it is not running all through the universe in some kind of a thread---and the same everywhere. As Einstein put it: "All that was needed was the insight that an auxiliary quantity introduced by H.A. Lorentz and denoted by him as 'local time' can be defined as 'time', pure and simple". (Abraham Pais, ibid, Ch. 6 & 7*.)*

Thence every time becomes somebody's local time. Therefore there are as many times as there are bodies in the universe. According to Abraham Pais, "There are as many times as there are inertial frames. That is the gist of the June paper's kinematic sections." (ibid, Ch.7.) To my mind this is the greatest of all discoveries (scientific, philosophical, logical, intellectual or whatever because time is the second most important thing in the world, second only to the mystery of life). It is simply phenomenal, the greatest of all intellectual ideas in so far as man is concerned, and we owe it to Albert Einstein. Basically, it means there is no such thing as 'the dawn or beginning of time' because there is not one overriding time to know its metaphysical origins and how it marches on and on, 'since the dawn of time'. This famous astronomical phrase has no meaning; astronomer will have to coin another phrase, for time did not begin with existence. Scientists decline to look at my work because it is 'philosophy', but I am afraid eventually they will be forced to do so. Einstein was not writing philosophy, yet he got the nature of time absolutely right and made it part of physics by showing that we can trace it to its metaphysical genesis—e.g. Lorentz created his t1 from scratch; and Einstein said it must be regarded as 'time, pure and simple'.

4.   So now we know that time can begin from thousands of places, and the duty of philosophers is to find out how each time begins logically, as the actual metaphysics of time overall. In other words, when we create our units of time, we create time metaphysically. The year is the beginning and end of all our time units--- and we created it. It is not in existence naturally. In the book cited below and given for free download on the web, I have speculated that any beings in the universe will have a time system roughly similar to our own, so as to accord with the physical reality spread out before them. It will have a large unit (like the year), subdivided down to smaller units for cultural purposes. All this applies to quantified time only---without which cultural use of time is not possible---as opposed to the silent passage of existence. Also, because of time, mathematics and a theory of numbers will exist everywhere in the universe where there are sentient beings. Thus a cultural system, science and technology, similar to what we have on this planet will permeate the cosmos of sentient beings. Only the quantum, I suspect, may be different, because the quantum is time-dependent, but time systems will differ from place to place.

5.   This supposition is consistent with the Einstein notion of time; and that is how important a notion it is.

Now the interpretations. What have the religious thinkers (in science and elsewhere) made of it? The local time idea began life technically as "The dilation of time as a measure of moving clocks". Its importance as the first occasion that time has been started by man in all history was missed by everybody (except Einstein). Instead it was a mysterious 'Time Dilation', meaning that time runs slowly with speed, which is an idea that seeks to make time even more mysterious than it has been in the Dark Ages. Yet there is no mystery in time dilation at all, since those travelling with the moving clock will notice no difference in its performance. Only those looking in from the outside will notice the difference. Einstein explained this with his theory of frames---the moving clock is in a different frame, a different world, if you like. Yet the religious among us claim that time slows with speed, and that, added to the Minkowski curved space-time, it means time will slow, or speed up and curve, so much that one could meet his grand parents even before they are born---harking back to the Ancient Greeks' Transmigration of Souls. When they add the twin and clock paradoxes to this, they conclude that time travel is real---'as predicted by Einstein's special theory of relativity'. I want the reader to know that this is rubbish, and that the people who peddle such ideas are not serious scientists, even though they may be clever mathematicians.

As I have said, the whole fault comes from one man, for it was not long after relativity was announced that Hermann Minkowski, a mathematician who actually taught Einstein at university (or The Polytechnic), came up with his own version of space-time. He said that by his 3+1 continuum, Einstein failed to show that time is mathematical, and proposed a new mathematical formula to make space-time geometrical, known as the 4-Dimensional Continuum, or 4-D Geometry (yet the Lorentz discovery that time can begin from anywhere destroyed his basic theory that time permeates all space and is the same thing as space, but the contradiction was not noticed). His basic equation is called ict, and he said it combined space and time into one entity. He was so vain that he announced that from the date of his theory time and space ceased to be separate entities---to which Professor A.N. Whitehead replied in a book (The Principle of Relativity), saying space and time still pass through nature separately. In Chapter IV of this book he wrote (and we have to remember that he was regarded as one of the greatest philosophers, mathematicians, and logicians of all time): "The heterogeneity of time from space arises from the difference in the character of passage in time from that of passage in space..."

6.   Even though the Minkowski equation rests on imaginary time, the reader must know that the whole of science from physics to Astronomy now uses the Minkowski formula, because, of course, Professor Eddington urged scientists to continue to 'employ it'.

If s=ct works at all, no doubt it is because (in our efforts to fool nature) earth time periodicities are carried in the head to the study of other metrics in violation of the Einstein theory of frames---and so long as we base all time on the earth-year, the units of time will always be the same, and therefore work for any system so structured.

But I don't think the Minkowski formula works well. In my opinion, this is part of the reason physics is experiencing difficulties---in the Strings Theory, The Quantum Theory, The Theory of Everything and other suppositions. It seems scientists are trying to fool nature with the Minkowski 4-D geometry: for if a theory is false it cannot accord congenially with the true physical nature of the world. I believe general relativity is in a mess because Einstein was eventually persuaded by mathematicians to adopt the Minkowski fiction for that aspect of his theory.

7.   The Minkowski theory was condemned by the great Bertrand Russell and Professor A. N Whitehead as arbitrary. As I have said, even the founder of Astrophysics, Professor Sir Arthur Eddington, the man who confirmed the general theory of relativity, called it arbitrary. He wrote (Mathematical Theory of Relativity, Cambridge, 1930, Ch.1): "...his space and time reckonings are imaginary surfaces drawn in the world like the lines of latitude and longitude drawn on the earth...Such a mesh-system is of great utility and convenience in describing phenomena, and we shall continue to employ it; but we must endeavour not to lose sight of its fictitious and arbitrary nature."

8.   Given this state of affairs, why is the Minkowski theory still used by scientists? My view is that it allows them to think religiously about time, since mathematicians are incurable mystics, much closer to theology than logic. On the contrary, the original Einstein notion of time leads to a completely secular time, originating from this earth, and limited to it. And yet time is obviously secular because somebody has to be there to place the points and count the orbits of the sun as 'years', and further break the year down to the seconds and so forth. Sentience is required; a theory of numbers is necessary; and the ability to count is absolutely essential. Above all, earth time is not part of the natural apparatus of the universe because it will disappear when the earth dies with the sun. Time on this planet is not connected to any other time anywhere; we use special postulates for our time on earth that are not known to exist anywhere else.

In addition, it makes time discrete, being the product of points that begins and ends on this planet. Discrete time cannot march; history is not 'The march of time' but the march of events. It is the long story of man's march in the world, being his continuing struggles with nature, and of many events, each of which is associated with its time of occurrence thus giving the dates of events; it is not the time that is 'marching on', but the events that are continuing and connected to their antecedents and consequences. Discrete time cannot bend or curve as in the concept of 'curved space-time' by which time travel is assumed to be feasible by all those religious scientists. Discrete time requires no 'arrow of time' for it to pass or continue. Discrete time can only pass by through the procession of its units---second, second, second; or minutes, hours, and even the years. There is only one year; we repeat it to get the centuries. This is the basic proof that time consists of units, separate and completely individual units, for we get all other units of time, short and long, as the subdivisions of the year. For example, we get the seconds from the year and call them units of time. But the year is only the physical orbit of the sun. Is the most profound inference about time clear to the reader yet? I hope it is, namely, time consists of the units of time; creating the units of time is the very act of creating our time metaphysically. Yet these units consist of mere physical space based on the space around the sun which we call one year; it is man who uses them as guides to action, so that one cycle means it is time to do so-and-so, etc---as stated above. We get the seconds from the year, but the year is only the space traversed around the sun. And the seconds are not set or chosen at random; rather they are divisions of the year chosen with such mathematical precision that a precise number (say 31, 536, 000, as I make them), equal one year---and start counting again for another year, on and on, to all the centuries for perpetual time. They are what we deliberately programme into the clock for reproduction not 'measurement'. Thus a time unit is equal to a specific amount of the space traversed by the earth round the sun perpetually, and we have memorised the categories (and organised our lives to accord with them) as the time units by which (alone) it is possible to live life satisfactorily on this planet.

9.   Logical consideration of these established facts leads to the solution of the problems of the passage and continuity of time, which has baffled the human mind since the beginning of life: the years increase in numbers to pass by together with their subdivisions all the way even to the cesium units or atomic clocks, since they have to be based on the second to make temporal sense.

Above all, the year is not naturally present in nature; we created it. And it is from the earth-year we have created all our other units of time, including the cesium units. Thus the definition of time as "the irreversible passage of experience", has been divided into specific periods. Definitely time is always passing; but we know of its passage through the duplication of its units because it has been quantified into specific units, known as years, months, days and hours, all the way to the seconds and cesium units. These we call 'time units', 'the units of time', or 'time in units', so that merely mentioning the word 'time' makes no sense. The quantity of time must de indicated, for time is known and used in units only. Hence the concept of 'quantified time' is very important. See APPENDIX I of my book cited at the end of this article; APPENDIX II is also very important, as it places logical restrictions on the powers of mathematicians. Without quantification we could never tell 'just how' time is passing, nor could we have it in convenient units for cultural use.

To repeat, we quantify time by linking specific amounts of space to the sense of duration in the mind. The sense of duration, of things and events lingering in the mind, is obviously the seat of the idea of the passage of time;

'during' implies a period, and that is time. Russell called time or space-time 'relation between points', and he was right. He described the relativity notion of time as "Perhaps the most important of all the novelties that Einstein introduced". And, furthermore, being a great thinker, the logical corollary did not escape him. Noting that the new Einstein theory abolished cosmic or universal time, he asked, "If cosmic time is abandoned, what is really measured by a clock...?" As usual, questions put by philosophers, if not their answers to them, tend to bring a subject to light illuminatingly, strikingly. For this question is one of the most profound ever asked about the world. What, indeed, is measured by the clock for us as time if it was not put there in nature as part of the evolution (or creation, if you like) of life?

10.   Unfortunately the question betrays a systemic, deep, and inherently confusing notion of time which makes it very misleading as a phenomenon. Time is closely associated with life, but what is it? We don't know, except that we age continuously, and experience (or existence as experienced), too, is always passing by. The best we can do is to create a system for telling "how much of it is passing by", and that is all the clock is manufactured to do for us---second, second, second, and so forth. We still don't know what time is. But with mathematics we can tell 'how much time' is passing (second, second, second), even without knowing what it is. The units of time constitute the time. In this sense, as I have argued, all we know of time is the quantity of it which we ourselves create with mathematics. That is what gives us 'limited time' for doing things in accordance with the relevant features of physical reality in order to be able to live satisfactorily in tune with nature; that is why time is what it is---namely, we can't do anything without it, and all because we know of the quantity of time passing. When it is midnight

you know you shouldn't go into the bush; when it is 7.am you know you shouldn't stay in bed if you want to keep your nine-to-five job, and so forth. By knowing how time is passing we can live in harmony with the features of physical reality; and we know this because time has been quantified to know how, or how much of it, is passing. 'Quantified passage' of experience is time, sentience is required for that; 'un-quantified passage' is bland existence such as happens to the inanimate matter in the world. That is not time; it is simply a matter of living, or existing, by means of physio/chemistry.

It is extremely important to understand this in the study of time. We are dominated by the passage of time; yet all we can know of it is how much time is passing. All life in the cosmos has to have a system for telling how time is passing so as to be able to organise life to accord with physical reality (the day and night system, for instance). That system or method is what we know as 'time', and it is obviously secular. But that is what has given us the notion of time in the first place. This must be clearly understood to unclothe the study of time of myths—otherwise it can be made so mysterious as to seem divine or magical.

After all, we cannot measure what we don't know; and the nature of time is unknown, as Russell's question implies. The nearest we can get to knowledge of the nature of time, logically, is using repetitive external mechanisms to tell us 'how much time' is passing. These mechanisms give us quantified time; so that the units of our time constitute our time metaphysically. The clock certainly indicates the amount of time passing in specific units; but it cannot indicate the true nature of time. All we can know is how much of it is passing---for time is always passing; it never stands still. We have carefully ensure that by choosing permanently repetitive cycles, or mechanisms; so that the units of time, by which we live, never dry up. The originators of our system of time (based on the earth-year) must be credited with the greatest intellectual discovery in human life---greater even than relativity. But they are connected. Einstein's train of thought is connected to time permanently, which is astonishing.

Thus to ask what is measured by the clock when you already have the clock is a confession that we (human beings) do not know the real nature of time—which is true. A unit of time is the metaphysical creation of *time*. For time is known and used only in units. But we create the units; therefore time is not naturally there in nature without the contribution of man. Man uses external mechanisms in association with the sense of duration in the mind to create his time units. The sense of duration is important because it is what distinguishes one unit from another, so that we know that one hour is shorter than one minute, and so forth.

In my system, time is always 'how much time'. In this sense, the clock does not measure time. Our method for telling how much time is the time. The clock cannot measure time out of thin air as if by magic. Let me stress that we do not measure time. We merely quantify it to know the 'quantity' of time passing.

This idea that we quantity time is crucial in our theory of time. The units of time constitute the time metaphysically. The quantification gives us the units, and the units are what we know as time, because all time is known and used only in units. When we say time is going we can only mean that its units (the years, for instance) are passing by, eventually to increase to a century; that is how time grows: second, second, second, all the way to the centuries. These time units are created elsewhere (with the earth-year and its logical or rational subdivisions consistent with physical reality), before they are programmed into the clock for 'reproduction'. So the clock can only reproduce the units of time---second, second, second, or minutes and so forth. The clock cannot measure time because no one knows what it is. Logically, it makes sense to describe time as the product of motion and chemistry---organic and physical. For what we call 'passage of existence' consists of motion and individual objects living or existing by their own constituent chemistry, which, put together, is called (collectively) 'the passage of existence'.

Unless this aspect of time is clearly understood, the new Einstein theory of time will itself be found to be inconclusive, or defective, so that we will continually ask 'what, therefore, is measured by the clock'? For otherwise where do the units of time come from? Where does the second come from? How does it manage to appear from the clock? The basic unit of time, the earth-year, does not even exist naturally in the universe. It is the mere physical orbit of the sun. It is man who finds it convenient to use it as a 'time unit', out of which all other units are derived with points, rational thought in consonance with physical reality, and mathematics. Thus all we can know of time is how it is passing by, or 'how much time?' by means of repetitive cycles in conjunction with the internal sense of time as duration: that is, the sense of things enduring or lingering in the mind. In the words of Professor Sir Arthur Eddington: "The rough measures of duration made by the internal time-sense are of little use for scientific purposes, and physics is accustomed to base time-reckoning on more precise external mechanisms." (The Mathematical Theory of Relativity, Ch. 1. Section 8.)

To put the same idea in mathematics, we quantify time with the aid of repetitive external cycles---the years, for example---in conjunction with the internal sense of periods (as units of time or duration in the mind), so that we can tell 'How much time'. Thus we know mentally that a second is shorter than a minute, and so forth. If you tell someone to wait for ten minutes you know mentally that it is not the same as requiring him or her to wait for ten hours, ten days, ten weeks or ten years. Our time units have been created with physical postulates found on earth, with the orbits of the sun for example; therefore earth time is not applicable to other parts of the universe in accordance with the Einstein theory of frames.

As Bertrand Russell put it: "There is no longer a universal time which can be applied without ambiguity to any part of the universe..." (ABC of Relativity, Ch. 5.) In other words, there is only earth-time as far as man is concerned, and it must be applied to the earth only. Carrying earth-time in the head to study other parts (segments) of the universe, even in the metric of general relativity, is not logically acceptable. How this earth-time came about and what essentially is its nature, are what we are trying to explain without the Minkowski fiction or theology. At least now we know why time is essentially discrete, for we can only start time with the application of points as applied to space—the year and its subdivisions, for instance, so as to have it in units. The irreversible passage of existence is divided with repetitive cycles, such as the years; so that, for example, one cycle means it is time to go to bed; two is to cook the food; three is to return from the farm, or go to school, etc. But culturally the year is so long as to be inconvenient for most activities. Hence we have learnt to subdivide it down to the seconds, or even the cesium units.

11. It is submitted that this is the philosophical definition of time consistent with the scientific facts as Lorentz and Einstein have discovered. However, from the way and manner we get the years, is it really wise to estimate the age of the entire universe in terms of the

earth-years? The sun is very young, we all agree. It took a very long time after the sun was created for the earth to come into existence, and longer still for it to support life. And are we saying that going round this tiny sun 13-15 billion times is the age of the universe, a universe of about a billion, billion major stars, some of which are so huge that a million of our petty sun can find room in them? Take a small hill and circle it fifteen billion times; can you use that to tell the age of the world, let alone that of the universe? And how do we define the temporal length of the year anyway? We cannot tell because the year is utterly indefinable in logic.

Let me stress again that we can never define the year on its own. For instance, when one year passes, you know that your life is shortened; part of your life is gone with it---but how long is that? The year cannot be defined in logic. However, we are sure it cannot be very long; and so 15 billion of that is certainly too short for the age of the universe. It is sheer humanoid vanity to estimate the age of the universe in years, I am afraid. Even then, of what value is that knowledge? Most of the cosmos should be left alone so as to concentrate resources for solving problems on earth, like tracing and neutralising the interstellar debris that threaten the earth, for instance.

Time, obviously, is a complex subject. It cannot be fully treated in A book of this size. Readers who are interested can freely download the book "Time in Science and Life: The Greatest Legacy of Albert Einstein" from my agent's website: www.practicalbooks.org---ISBN 13: 978-1-4092-6809-3. The whole contents of the book are put out free, since money is not the object in a debate of this nature.

Time And Quantified Time

We are all fond of using the word 'time' loosely to refer to the passage of existence in any form whatsoever. That may be called 'the unscientific' notion of time. In logic, science and philosophy, however, time is what Professor Richard Feynman called 'how long we wait'. This translates into the concept of 'how much time', or quantified time, so as to be able to tell how long we wait in mathematical language for universal application.

In a serious discussion of time, it does not make sense to just mention time. The context of any proposition (in science, mathematics and philosophy) must always show or imply the sense of 'how much time' in it, or expressly show the quantity of time proposed. Of course, time may pass when one is not conscious of it. But in all cases, when one wants to know how much time has passed, or will pass (as in futuristic propositions), mathematics must be used to quantify the time. And let me stress again that we quantify time by the use of external cycles in union with any sense of duration of anything whatsoever.

Quantified time is 'time in a clock', any clock at all. And the clock, any clock, can only show time as independent of space. Space-time is automatically quantified as it is derived from space with points, which is the only reason for calling it 'space-time'. Discrete time can only pass through the succession of the individual units. On this point, Leibniz was absolutely right when he said time is succession. What was lacking in his day was the concept of discrete time; with this new concept in our post-relativity world, we can now see clearly as to how time passes and seem continuous through the succession of its separate and individual units: second, second, second; or minute after minute after minute. Plus the hours, weeks and months all the way to the year, which also passes in the form of year after year after year.

It may seem surprising, the springs of a thousand legends, giving rise to supernatural speculations, that we have an extremely ingenuously smooth time system, so cleverly structured that it is there when we are born and there as we die, and always passing by. For this reason we know that "Time does not wait for anybody", not even Kings and Queens and Presidents. Even surrendering one's Kingdom and all possessions for a moment of time cannot save the most powerful Queen on earth. Scrutinised under a logical gaze, however, time is not so rosy; it is only one moment, repeated to pass by and seem continuous so that arithmetic can be applied to its accumulations. [7] This, as we know well, happens when we reckon time for futuristic planning, and backwards as history.

---

[7] Let me explain that space-time is necessarily discrete. We have only recently come to understand space-time from Albert Einstein; yet time has always been discrete, consisting of only one unit (or moment) of time---of whatever length. For there is only one year, and all other units are obtained from the year in the form of separate units of time. To get more years we repeat the one year exactly. Thus we have second, second, second; or minute, minute and hours and so forth. Each is a moment (or a unit) of time in its own right. The notion is best illustrated with mechanical devices: if you set a timer to regulate the working of any mechanical device, when the time ends, the machine will stop because the time allowed has ended---time, even in this sense, has a 'beginning and end' based on human activity, as Lorentz discovered, and Einstein was right to extend the notion to all time, or time *per se*, exactly as he put it, 'time, pure and simple'.

But for the union between the sense of duration and external cycles giving us units of time out of the moments of time, time for the clock would not exist at all.   Presently philosophers see time as rather a straightforward pragmatic entity, albeit not as simple as it is normally supposed. It is partly a confidence trick, which makes the clock work continuously, the trick of continuity is in the repetitions of the seconds, or of the units of time, all of which are to be understood as single moments---which are the realities--- of quantified time. It is also partly physical (using physical cycles for the process of quantification); and partly philosophical, i.e. according to Einstein without time physical reality is indecipherable, or cannot be properly (accurately) determined.

What Is Measured By The Clock?

Our time is based on the repetitive orbits of the sun by the earth, and evidently the earth never stands still. If ever it does stop going round the sun, our time system will be completely nullified; but, of course, life will go on. It is inconceivable that all life will be extinguished instantly the moment our time is (mathematically) nullified in the sense that only quantified time would be lost. This is the best proof there is that life is not based on "time allowed", as the religions believe; rather time is a union between the sense of duration and external cycles---therefore man had something to do with the time we have in the clock, the only reliable time, as quantified time.

All the religions speak of "time allowed" for the duration of a man's life. They had to, because the nature of time is easier to explain as a providential bounty than anything else. To be honest, without a cosmic explanation for time, what is time, or, put the question in another form, what is the origins and essential nature of time? Of course it is assumed that the clock measures time---but from where? And what it is that the clock measures? [8] The clock maker will say he invented the

---

[8] Without the explanation that what the clock measures are cycles of duration, or duration reduced (or converted) to cycles repetitive, metaphysically interpreted as a union between duration and its conversion to external cycles, time can never be logically accounted for. We will just go on using it---but in what form? In the form of units (year after year after year, and all the seconds and so forth derived from the year); yet that means the same thing, namely, a union between duration and its conversion to external cycles. For the year is only a cycle of the sun. It is not time. It is the practice of humankind to call it 'a year; and we use it as our basic unit of time, as a matter of convenience. Otherwise in nature it is not

clock to reckon time in the sense that everybody knows---but what is that sense of time?

When it is postulated that general time permeating the whole cosmos (and therefore the same everywhere) does not exist, the first implication is that every body has to have its own time; it is not coming from the cosmos therefore it must have originated on this planet.   So let's find out how it all began. That is the first implication. The second is that, as a result, cosmic time is abolished---although it sounds tautological, it still has to be emphasised, as well, and most clearly because the 'cosmic time instinct' is permanently ingrained in the human mind.   One reason is that time cannot be suspended; but the more cogent reason is sheer intellectual incompetence plus fear of the unknown. We are always using it, and so it does not make sense to just insist that it is not there. If it is there, and did not come from the cosmos, how did it begin?   And the obvious fact is that it is always there. Even before we are born, and also as we die to leave it behind.   Yet it cannot be supposed that each body's time is a version of something 'naturally existing', whether it permeates the whole cosmos or not, with the necessary but illogical (little 'academic') proviso that it may not be the same everywhere but varies with individual bodies in accordance with unknown natural laws.

It is plainly evident that this erroneous sense of time dominates scientific thought. Hence time is not defined in physics; and as a result, the Minkowski fiction makes sense to some scientists, including even Albert Einstein himself. So far only Professor Arthur Eddington has redeemed physics by warning that it must never be

---

time. As a matter of fact, we can use something else---we can tap the finger, for instance.

forgotten that the Minkowski formula is "fictitious and arbitrary"---but they have chosen to ignore him.

Thus Russell's query is important, namely, "If cosmic time is abandoned, what is really measured by a clock...?" [9] My answer, of course, is that outside the union between the sense of duration and its conversion to external cycles, time does not exist to be measured. The very act of 'measuring' is the time in essence---like moving from point to another point, time is going, so that time becomes 'relation between points'. The cycles are time units (the years, for instance), and the time units constitute the time: a year is a cycle, but it is our time, the basic unit out of which all other units are derived.

However, the cycles are the creation of man for the sole purpose of converting the sense of duration (of any thing or any event, like the period it will take to reach the village from the farm before nightfall to avoid predators), to his time units to guide his activities. So the clock does not measure time; it rather reproduces units of time programmed into it repetitively---second, second, second, and so forth. It should be remembered that the seconds are put there by the clockmaker; but where do they come from? The answer is that they come from the subdivisions of the year. Otherwise the time does not exist anywhere to be measured---the units constitute the time. Without the year there will be no seconds, and the like, all of which are derived as subdivisions of the year. As hinted above, you can even dispense with the year and its subdivisions and tap your finger, if you will not get tired. A million taps means it is time to go to bed, and so forth; outside the units of time, time does not exist to be measured; but the units are the creations of man as quantified time to record the

---

[9] ABC of Relativity, Ch. 4.

passage of existence in his experience in manageable units for cultural purposes.

The Principle Of Mathematical Equivalence

In nature there is reality and our perception of it. In the word 'perception' everything man does in life is implied, including mathematics, since we can only act by perceiving the true nature of the physical world; I am using the word in a sense akin to 'experience'. The problem is, mathematicians normally are permitted to imagine things to satisfy their nostrums, so that they do not rely solely on their percept; however outrageous, they can defy reality, logic and common sense, and leave it to the applied mathematicians (physicists, astronomers and cosmologists) to find out whether what they have assumed is there in nature so that their theories based on it can be seen as true or not. In no other profession is this sort of thing allowed. Even one of the greatest mathematicians Britain has ever produced, Professor Sir Arthur Eddington, criticised that common mathematical tendency in his book, The Mathematical Theory of Relativity. I have quoted him above in the text, but it will do no harm to repeat it as it is vitally relevant here. He said: "The pure mathematician deals with ideal quantities defined as having the properties which he deliberately assigns to them. But in an experimental science we have to discover properties not to assign them..." The principle of mathematical equivalence should make them think of the practical consequences of their imaginary properties, although I doubt it, but that is another matter. The rule is that mathematics should not seek to make the basic features of nature what they are not quantitatively; any such propositions are bound to falter. Note that we are talking only of basic phenomena. By the very nature of man, it seems he can make qualitative changes in peripheral nature not quantitative changes in the fundamental aspects of nature, and time is the second most fundamental feature of both nature and life.

The principle means that, in effect, one cannot use mathematics (sometimes defying comprehension) to state, say, that there are ten trees in a field, and propound theories about them if, in actual fact, there are only two. This is slightly different from assigning imaginary properties to nature. It is different because it relates to 'quantities'. Six into four won't go, or something like that. The principle of mathematical equivalence rules that, to accord with physical reality, one can only talk about two trees, or as things are not as the mathematicians want them to be. Nature is not there for the convenience of mathematicians; it is neutral. That was the advantage we gained when the ancient teleological interpretations of phenomena was discredited. Therefore this rule is not to be scoffed at. I regard it as one of the strictest doctrines in logic and metaphysics.

It is not often realised how progressive is the study of philosophy. Quietly but surely, many entrenched myths from our primitive past are being discredited one by one by philosophers. One of them is teleological argument. With that and many other ludicrous intellectual fashions out of the way, it is unacceptable to regard any concept as 'compounded for the convenience of the mathematician', as Russell defined the Minkowski theory of space-time. Some day, we may get scholars writing about the many myths philosophers have discredited through their quiet researches to foster science and progress generally. So I regard this principle of mathematical equivalence as a strict and necessary doctrine to prevent mathematicians arrogating the power and right to alter nature quantitatively in the fundamentals of physical reality. We shall, and should, continue to alter nature qualitatively to our benefit---

gardens, buildings, roads, cities, waterways, canals, railways, bridges, tunnels, all science (bar destructive devices), and all art, sports and so forth. They do not change nature but beautify it; but quantitatively, never. We cannot make one object two, or two objects one, physically. It is not possible realistically. Not in reality only in the imagination. The only one I know of that has achieved any kind of academic adherence in the strictly rational post-relativity era is the Minkowski formula, but then it is regarded as fictitious. Thus mathematicians who rely on it must know that they are falsifying their nostrums.

The origin of the rule will help the reader to understand it well when spelt out: it occurred to me when I was pondering Hermann Minkowski's claim to have made time and space into one entity as from the moment he outlined his theory, as previously quoted, in the following outrageous (even cheeky) statement: "The views of space and time which I wish to lay before you have sprung from the soil of experimental physics, and therein lies their strength. They are radical. Henceforth [that is, from the moment of his lecture] space by itself, and time by itself, are doomed to fade away into mere shadows, and only a kind of union of the two will preserve an independent reality". This is to make two things in nature into one with mathematics ('a kind of union of the two...') So he knew they were two independent things. How could he have made them one from the very moment of his lecture? He spoke of experimental physics. In fact, the only experimental evidence pointed to time being 'local' in nature; and Einstein adopted it in his special relativity. There was no suggestion that time had been found to be inextricably intertwined with space---rather the suggestion was that time could not be had without space; and that once you have space, you can create your own local time. What Einstein did was to interpret local time to mean "The only Time" we can have.

The actual physical reality known to be in existence was precisely as Minkowski himself stated it---namely, that time and space were two separate things. But it is interesting that he sought refuge in experimental physics. In that sense he did not breach the principle of mathematical equivalence. It shows that he was really a very good thinker; he had to be that good to convince Einstein to adopt his formula for general relativity, which came ten years later. The unfortunate thing for him is that the evidence he cited was really irrelevant to the claim he was making. He needed physical support that time and space are inextricably intertwined and therefore constitute one entity. The evidence that had been discovered by Lorentz and Einstein was that time was essentially local in nature, leading to the supposition that 'there are as many times as there are bodies', and that, additionally, time is different in different places, and also under different conditions. The principle of mathematical equivalence can be used to refute Minkowski's claim to have made them into one entity as from the moment of his lecture.

The rule stipulates that he could only have spoken about time and space as they actually were in physical reality, which, he has admitted, were two separate entities. The reality before Minkowski was that there was space, and there was time. Even the great Einstein himself made them independent in his special theory of relativity. So it did not surprise me that Professor Sir Arthur Eddington and Bertrand Russell described the Minkowski proposal as arbitrary and fictitious. However, it did surprise me that mathematicians ignored this strong condemnation to claim that they could not understand Einstein's ideas without the Minkowski fiction.

This made me sit up and think, think of a principle to require mathematicians to relate their suppositions to exactly the nature of physical reality laid out before them, not as they would wish it to be to accord with their nostrums. The result is that I came to the conclusion that mathematics can only mirror reality, not to alter it with mathematics alone. So the principle of mathematical equivalence is this: Mathematical statements (or equations) must strictly accord with physical reality. It means no mathematical quantity can exceed or reduce what the actual physical quantity is. No mathematics can make one thing two, or two things one, without physical divisions and unions. As applied to Minkowski, he failed because, as Professor A.N. Whitehead has pointed out, time and space still pass through nature as two entities, not one. He could not achieve the physical union---it was, alas, only mental, imaginary.

Why Space On Its Own Is Not "Space-Time"

In Einstein's special theory of relativity, we learn that, "In the absence of gravity, space and time are distinct entities. In the metric of special relativity they play distinctive roles." [10] Nothing in special relativity has changed since then to make all space "space-time". Yet in all their suppositions cosmologists and astronomers always refer to space as space-time.

Let me set out the facts as they are at present, as argued all through this book, and hope they will see the light. To begin from the proper beginning, the whole idea of space-time comes from H.A. Lorentz; until then space was space and time was time. It is true that in special relativity Einstein made space and time dynamic rather than the Newtonian absolute; but being dynamic merely means they are changeable under different conditions. But about time alone Einstein avers that he was able to complete special theory of relativity five weeks after he gained the insight that the Lorentz idea of 'local time' can be defined as 'time, pure and simple'. So let us examine the Lorentz notion of local time.

H.A. Lorentz found that time runs slower when in motion, known as "the dilation of time as a measure of moving clocks". He could not understand why and literally put it aside. He called it 'local time' or t1. To him it was not 'the true time' but a mathematical auxiliary or curiosity--- not very important. Time, he said, was time, denoted with t, and t1 was something you get as your local time,

---

[10] Professor Jeremy Bernstein, in *ALBERT EINSTEIN: and The Frontiers of Physics*, Op. Cit. p110

but certainly not applicable in the outside world as time, because it was a mere mathematical curiosity. May I remind the reader that all this has been given in detail in the text above. I have even mentioned Lorentz's own statement that he thought he failed to discover special relativity because he did not regard time dilation as of any importance.

Strangely, however, as one of his brain waves, Einstein worked this into his theory of frames. The dilated time was 'local time'---the time of your locality. Now, if the universe was fragmented, then local time would be somebody's time, which to him would be running normally like any other time, but to outsiders, would be running erratically (or slowly, in this case.) In actual fact, that was the case with the Lorentz discovery. People outside the moving clock would see it as running slowly; but those carrying it in the moving vehicle would notice no difference in its performance. That is the genesis of the Einstein theory of frames. Otherwise time was separate from space. What you will find is that it varies under different conditions, simply because everybody has to have his own 'local time' in his locality or inertial frame. But since time is continuous, and having made it a separate co-ordinate in the study of phenomena, dynamic space would have different time co-ordinates at every turn. We recall that Bertrand Russell has stated that from the sun's point of view the tram never repeats a former journey---because the time co-ordinates would be different. Since time is a separate co-ordinate in the determination of physical reality, different time co-ordinate implies a different situation, different physical reality.

This was the state of affairs when Hermann Minkowski came in with his theory of 4-D geometry making time part and parcel of space---all space. So that cosmologists and astronomers call his theory "The

Minkowski Universe", meaning that all nature is subject to the 4-D geometry, where time and space constitute one entity. But let us swiftly add that the foremost mathematical interpreter of relativity was our own Professor Sir Arthur Eddington, the man who confirmed the general theory of relativity. He wrote the definitive book on relativity, called The Mathematical Theory of Relativity. About the Minkowski 4-D Geometry, he stated clearly on Page 9 (Ch. 1.1.), as already quoted, "Such a mesh-system is of great utility and convenience in describing phenomena, and we shall continue to employ it; but we must endeavour not to lose sight of its fictitious and arbitrary nature." [11] He was not the only great mathematician who described the Minkowski formula as arbitrary. Bertrand Russell also said it was based on arbitrary assumption. As quoted in the book, he made it plain that because of that the derivation of the Minkowski 'interval' as time from space was illogical, or invalid.

Let me try and explain again the reason mathematicians still adore the Minkowski theory---even though they know that it is fictitious. It makes things easy for them. Yet it is not true. They accept the novel Einstein notion that time must be made a distinct co-ordinate in the description of phenomena. You see, the problem is that Einstein made all time (any sort of time) 'local time'---the time you create for your own local purposes, as Lorentz had discovered. Einstein extended the Lorentz idea to all nature. With the universe being fragmented, it was impossible that one system of

---

[11] The emphasis is mine. I have had to mention this several times, because, quite honestly, I am outraged by the mathematicians' desire to perpetuate the Minkowski formula as if it is really true of physical reality---yet it is not, and they know it. At least one of their own numbers told them so.

'dynamic time' (as opposed to 'absolute time'), would apply with equal validity to all fragments of the universe. As a result he said there are as many times as there are bodies in the universe. Nobody can contradict Einstein on this matter. But mathematicians found that creating your own time to add to phenomena to acquire your concepts of physical reality puts too much power in the hands of mankind. (I suspect there are religious sentiments in this.) [12] Besides, it was complicated. The Minkowski system was easier; [13] you just have to mention the Minkowski space or ds2 and move on. It comes with time already embedded in space as part of it---so the whole of space is 'space-time' and every time is also 'space-time'. The caveat of Professor Eddington was quietly ignored. Soon everybody forgot about this; Eddington and Bertrand Russell were dead; and there was nobody clever enough to notice the discrepancy and question them about it. Of course, that leads to a distortion of relativity, but mathematicians are the arbiters of truth in mathematical physics and they were

---

[12] The Minkowski formula makes time universal again after Einstein, namely as (s=ct); something in general existence mysteriously (harking back to Pythagorean mysticism in mathematics), which can be invoked with the appropriate mathematical symbols; not as something you create in your own local space with the application of points to space, which makes time completely secular. It seems to me that humankind is not ready to accept time as purely secular. Those of us who have already made the necessary psychological adjustments for accepting time as plainly secular are not regarded as normal.

[13] It was difficult in mathematics but easy in logic and philosophy; and let me hurry to add that, because of the involvement of time, the whole notion of local time or space-time has philosophical implications, since time is the second most important thing in the world, second only to life itself.

the ones benefiting from the Minkowski theory, and therefore preserved it. Otherwise it is not true that all space is 'space-time', while all time is also 'space-time'.

Yet it is true that time is always space time. You cannot have time without space; not because the space comes with time inside already, but because all time is known and used in units and units only, which can only be had by the application of points to space. There are elements of time in the mind as the internal sense of time, known as the sense of duration. But we have got to link duration to external cycles to give us usable time in units, as I have explained above. For example, without space we cannot have the year; yet the year is our basic unit of time out of which all other units are derived. This brings a little complication but nothing serious. The reason is because you can only create time, as 'intervals', or as 'time units', as I suppose (because the year is only one unit of time and we derive all other units from the sub-divisions of the year with points or mathematics), with the application of points to space, thus making time a product of space, and therefore 'space-time'. The truth of the matter is that you cannot have time without using points to divide space; it makes time necessarily discrete, being the product of points. Therefore time is always 'space-time, or properly 'space-timed'. But that is all the connection between space and time, except that space is required, again, for displaying time in units as we have in the clock. [14] The

_____

[14] As discussed above, the poignant question posed by Bertrand Russell comes up again, namely, in the absence of universal time, what really is measured by the clock? (ABC of Relativity, Ch.4.) This is a very serious matter, because if cosmic time is abandoned, there is no time, or any logical explanation for the time we have. The answer, of course, is that the clock does not measure time. It is deliberately programmed to *reproduce* specific units of time: second,

clock, any clock, does not give 'flowing time'. It merely reproduces units of time programmed into it. The old mechanical clock based on coiled springs gave the best illustration. The springs are manufactured to release units of time: second, second, second. If one failed to rewind the springs, the clock stopped ticking. The springs provided the clock's energy, but were strictly programmed to reproduce time in specific units only.

After the time is derived in this way, it becomes separate from both the space and the points used in creating it. That is why Einstein made them separate entities in special relativity. For, apart from the condemnation of the Minkowski 4-D geometry which assumes that time and space constitute one entity by Russell and Eddington, Professor A. N. Whitehead has also pointed out that time and space still pass through nature separately---not as one entity. To add to these, I have humbly suggested the Principle of Mathematical Equivalence above, which can also be used to denounce the Minkowski arbitrary and fictitious formula.

---

second, second, leading to minutes and so forth, to accord with the cycles of the earth, so that about 31,536,000 (or so many) seconds will coincide exactly with the earth's orbit of the sun, called 'one year'. To have more years, we go round the sun again and again and again---hence perpetual time. Units of time in procession give us continuous time. From the Einstein concept of space-time we know that time, since it is produced with points, has got to be wholly discrete.

THE MISCONCEPTIONS OF TIME IN RELATIVITY

It must not be supposed that the problem of time in relativity has been conclusively settled. Relativity is physics. When a problem is solved in physics the solution is always clear, precise in mathematics, and universally applicable; but time in relativity at present is very vague, neither definite nor precise, not least because consideration of time is a philosophical enterprise. The argument is that the original Einstein theory of time can be used to solve the passage and continuity of time. Unfortunately, Herman Minkowski made the question of time in relativity immensely complex and vague, not at all like the original notion proposed by Einstein. Indeed, as a result, the question of time on the whole is destine to keep the philosophers busy for several centuries as their nostrums become footnotes to Einstein instead of Plato. As regards the physicists and cosmologists, as opposed to the philosophers, they believe that the Minkowski theory makes things easy for them; the problem is that it is just not true of the physical world.

Bertrand Russell has said the concept of space-time is perhaps the most important theory Einstein introduced. To me, there is no doubt (no 'perhaps') about it. It is the most revolutionary theory in human history simply because time is second in importance only to life itself--- and yet that life cannot even be lived as a well-organised existence without time. That is how momentous time is in human affairs; and Einstein has shown that it is very different from what it has been traditionally assumed to be. Secondly, he insisted that it should be taken as a separate coordinate in the study of phenomena. In the determination of physical reality, because of Einstein time is a co-ordinate in its own right just like the height or length of matter and space are, thus making Man, the observer, part of the observed, since he has to add the time in the 3+1 formula. Those mathematicians who assume, on the Minkowski theory, that time can be incorporated into space with mere mathematics so that we can dispense with the 3+1 formula and the metaphysical role of man in the determination of physical reality, are contradicting Einstein, which is something approaching a hanging offence in science. On the contrary, it is possible that the passage and continuity of time can be conclusively resolved with the original Einstein theory of time as space-time, or local time.

There is obviously fear in some people that time cannot be something we invent by ourselves. Of course, if 'there is no longer a universal time' we have to find out how we get our time. [15] However, nobody is claiming that man invented the whole of time. Rather we have found that we invented how to quantify time by linking the natural sense of time as duration in the mind to external cycles. This sense of duration of anything is obviously connected with the memory mechanism for the retention of images and concepts in the mind.

Let me stress again, and more strongly, that the sense of time is duration in the mind. In his Mathematical Theory

---

[15] It is not often realised that philosophy is of great importance to science; and, as an example, this is the sort of thing philosophers do behind the scenes to make their suppositions indispensable to science in general; for the philosophers service every branch of science. The phrase 'survival of the fittest' from biology which has passed into general usage in science and linguistics, was coined by a philosopher, not Darwin. All the sciences need philosophical interpretations. In the quotation above from Professor Dingle, he was saying this very strongly in respect of physics; but all the sciences need the same sort of assistance from philosophy, including even mathematics and logic.

of Relativity, Professor Eddington made this absolutely clear, as quoted above; and we have got to take that view seriously because the theory of time outlined in this book is based on relativity. Unfortunately the mental sense of duration is not enough. It cannot give time for general use because it is private. The word 'time' is meaningless until it is objectively quantified. We need time in units to apply to the external world---i.e. to mechanise in the clock for general use, so as to be able to tell 'How much time' at a glance---see Appendix I above. This is achieved with external cycles, the most basic of which is the earth-year out of which all other units of time are derived with mathematics. And it is maintained that this is in complete conformity with the Einstein notion of time, and therefore incontrovertible. Above all, it is the only means by which we can logically solve the problems of the passage and continuity of time.

For now, we are told in all earnestness from the discussions above that relativity is not properly understood. This may be so. But actually relativity is only a theoretical system, a suggestion. It is based on the suggestion that physical reality is not homogeneous but fragmented, and therefore subject to different natural laws. This applies to both special and general relativity. Bertrand Russell called it 'a logically deductive system'. In plain language, 'a new philosophy of physical reality' so logically structured that it demands attention, respect and serious study. And these Einstein has certainly achieved. With Einstein alone we are not talking about genius but a godlike intellectual phenomenon never seen on this planet before; he reconstructed the world of physical reality single-handed, that is the reason he is indispensable to both scientists and philosophers.

So Bertrand Russell was absolutely right. Einstein's system is a new logic of physical reality, and it works. But theoretical physics is most unlike the physics we apply in laboratories. Ordinary physics is much more like chemistry; it has consequences. The Nobel Committee was right to award Lord Rutherford the Prize for Chemistry, even though he regarded himself as a physicist, who had rather cheekily claimed that "all of science is either physics or stamp collecting"!

In theoretical physics there are no obvious consequences, so it is difficult to judge the merits of suggestions. Instead, when we get a new theory in advanced physics (rightly or wrongly), three things will happen. I mean, all three will definitely happen in succession, whatever may be the merits of the new proposal. First, we will get interpretations of the basic postulates proposed in such complex settings (or confused formats) from rival theorists that the debate just has to go on; nothing will be settled in the meantime. But because there are no consequences, nobody will get hurt, no machinery will fail to function; avoidable calamities will not occur. The rains will not stop; the sun will not dim.

The most recent example was the eather debacle (or debate). Secondly, we will get accusations and counter accusations of misrepresentations and misunderstandings. The third possibility (because philosophers share with theoretical physic one subject-matter, being the determination of physical reality), will be philosophical interpretations to arrogate the almighty right to shame and discredit some of the factions in the debate, only for philosophers of different schools to turn the tables---and so the debate will be carried on and on. These philosophical discourses are often pretty profound, giving several intelligent interpretations without being able to settle the argument one way or the other. Strangely, that is how we eventually acquire our knowledge of the external world, sometimes referred to as the practice of 'academic freedom'. That is what happened to Plato. And that is what is happening to Einstein as he has come to replace Plato, in fact, to make his basic suggestion redundant, if not completely false, due to the quantum theory.

A careful examination of what has happened to Einstein's theory of time so far betrays elements of all three conditions. First, we are told that 'most definitely' due to Einstein's analysis of 'Order and Simultaneity' there simply is no 'standard or absolute time frame in the universe'. ('Time Frame' or 'Time Reference' means the same thing. It means the logical criterion of validity.) This is generally accepted as true; for it is reinforced by the Lorentz time dilation and local time concepts.

However, it implies that time in the abstract is utterly indefinable, as I have shown above with discussions about the earth-year. The year is indefinable; other time units in use on earth are defined in reference to the year. But the year on its own is logically indefinable. Again, all our time units, down even to the cesium units, are based on the earth-year; they are meaningful only as related to the year; but like the year, on their own (that is in the abstract) none of them can be logically defined. How long, for instance, is a second in logic without reference to something else? The result is that we all have to use the clock, or clocks, based on the earth-year. By this theory of time (as quantified time), the human intellect is built upon the concept of "points and instants". Instants do not exist independently in nature. Only points do; they had to be discovered by man, but they do exist in nature independently—for example, trees constitute points. Before we learnt to put points on paper, we could see that trees dotted the landscape. Thus points constitute the basic instrument of human thought, especially in mathematics from which all science spring. The instants arise from the act of 'consciously' and 'purposely' moving from point to point, confirming the Russellian notion that time is 'relation between points. Hence quantified time is human in origin, except that the internal sense of time (as duration of anything in the mind) must be recognised as making a psychological contribution to the invention of quantified time in that the external cycles used for quantified time (the years, for instance), have to have psychological anchors (meanings) which are the senses of duration of anything in the mind.

Secondly, in the absence of a standard time frame, what does it mean to claim that time intervals in a moving frame are shorter---shorter as against what kind of standard or universal time? What time intervals are they compared with since there is no standard time frame? (Note that you cannot say they are shorter as compared to other clocks outside the moving frame; that will bring in the Einstein theory of frames, as I will discuss presently.)

So we all, in the end, have to resort to using the clock or clocks based on the earth-year. Yet if we use the clocks then it is not correct to claim that time intervals in a moving frame are shorter; they are not naturally or normally (in its proper setting) shorter or longer; they are normal to that frame, or to its natural frame. The moving clock may only seem 'different' as viewed from the outside; but if that is the case then there is no puzzle. [16]

---

[16] Otherwise it is difficult to see how the behaviour of one clock can affect all time, human physiology and even the material contents of atoms, e.g. muons. If time is defined as the passage of existence in consciousness, how can the behaviour of one clock affect it for all of us? There is still a lot of religious beliefs about time. Time dilation is one of them, so sweet to the religious in science because they can claim that "it is a unique mystery about time predicted by Einstein". In

The time of the moving frame is not 'our' time; and it is not queer to its natural environment or setting. It is a strange phenomenon to those looking in from the outside, in breach of the Einstein theory of frames. In fact, it is irrelevant to anybody but those in the moving vehicle only.

The whole idea of studying other frames from the outside is fraught with difficulties; it can never be an exact science since the standard postulates that make our system work (and make it what it is) might be inapplicable outside our frame, or planet. [17] Speculations into other frames from our frame have been responsible for all the bizarre suppositions about time and space-time from mathematicians and cosmologists in general relativity. I don't think that kind of enterprise is justifiable, especially when it leads to theories that space-time may be infinite in its timelike directions. Space-time cannot be infinite because it is necessarily discrete---the year, for instance, is not infinite. It is only one; all other units of time derived from the year are also discrete and individual. The proper way to think of time as space-time is that its units are in perpetual procession (one year or second following another) to

---

fact, it is not a mystery, let alone predicted by Einstein: he rather solved the little problem with his theory of frames—i.e. the dilated clock belongs to another frame to which it is running normally.

[17] I think one implication of this is that the laws of physics, or some of them, would differ from ours at least in some parts of the cosmos, if not all over. Einstein was really a very strange genius in physical thought. He introduced the notion of postulates for natural laws in frames. This idea may go very far indeed in the cosmos at large.

make time seem continuous; as such time can never be infinite.

Nothing illustrates the confusion about time in physics as a result of relativity and how it is misunderstood by scientists than the story of muons. By normal logic they should not last long enough to reach the earth; but they do. With the use of formulaic mathematics and concepts, physicists explain this by saying special relativity provides the answer as follows: the speed of muons is so great that their internal clocks slow down. Using the theories of time dilation and the so-called twin paradox based on it, it is assumed that as the muons speed and their internal clocks slowed down they age less and thus are able to last long enough to reach the earth. To a logician or philosopher who understands relativity, this is so laughable as to choke him. It is really the best example of the confusion in physics about time in relativity. (1) Time dilation has nothing to do with the muons and how they behave, since time does not dilate internally. Lorentz found that a moving clock would be seen by outsiders as running slowly; but internally those carrying the moving clock would notice absolutely no difference in its performance. Einstein explained this with his theory of frames---the moving clock is in a different frame. There is no logical mechanism for this kind of episode to be able to control time per se. All other clocks would not run slower or faster; and since there is no such thing as absolute time frame, or a standard time, by which all other clocks can be compared, the moving clock's performance has no relevance at all in physics, because its carriers would notice no anomaly; and those outside who notice any anomaly should mind their own business since it is not their time. (2) The idea that muons have internal clocks is based on the Minkowski theory of space-time, where space and time are assumed to constitute one entity; and therefore the reasoning goes that, since the muons occupy space, and all space is space-time, they have their own internal clocks to keep or measure time for them. Again, any logician will describe this as nonsense; for after all, the Minkowski space is known to be fictitious and arbitrary with absolutely no logical validity. Secondly, the very idea is easily disproved thus: we know there are (roughly accurately) specific times on our normal clock for certain events on this planet---let us use Sunrise and Sunset for illustration. If Sunrise is usually 6 am and Sunset is roughly 6 pm as they are in some countries in the Tropics, it is inconceivable that a moving clock can force or influence these times to become 5 am and 5 pm, on the planet just because it is running an hour late---simple.

The reader will have noticed that the name of Lord Bertrand Russell comes up regularly in all discussions of relativity's interpretation. It is inevitable. Russell was highly respected by Einstein, and for very good reasons. He was the greatest philosopher of the time. He was also a great mathematician and logician of genius. A most attractive writer, who won the Nobel Prize for Literature, he wrote about every subject in philosophy, including novels to illustrate moral points. When relativity was announced, he abandoned many of his most cherished ideas as wrong without shame or even mild embarrassment. He was candid and honest in the most adorable way, completely dedicated to the truth no matter how it reflected on his own beliefs. Russell probably had no certain beliefs other than the pursuit of the truth wherever it took him: via science, logic, mathematics or plain common sense, and linguistics. If he was certain that teaching mathematics to people from the cradle could save the world, he would have advocated that as his philosophy.

Concerning relativity specifically, in the later editions of his little book "Problems of Philosophy" he denounced his original philosophy as expressed in the book because of Einstein's theories, joking that whoever wrote the original ideas must have been a monkey, but nobody should suppose that the monkey looked, even remotely, like himself! No great philosopher has ever made such a confession; often associated with rulers, they all wrote imperious edicts as if they had discovered the final truth in logic and metaphysics. [18] Indeed, Russell later called his Fellowship dissertation "somewhat foolish" for the same reason, namely, the geometry used by Einstein had made his discussions of the foundations of geometry completely wrong, and he was happy to admit it and adopt the new Einstein theory. He wrote one of the best interpretations of relativity, still in use, under the title "ABC of Relativity". His book "The Analysis of Matter" can be divided into two. One section is about relativity; the other is mainly about his joint theory

---

[18] No surprise, then, that Russell later put them in their deserved places (mostly of dishonour) in his monumental *History of Western Philosophy*. One complaint is that he never even once mentioned the name of Wittgenstein in this great book. The reason came from his contemporary, Sir Karl Popper---it was because, "In the long history of philosophy there are many more philosophical arguments of which I feel ashamed than philosophical arguments of which I am proud...Russell saw these things in that light, and so did I..." (From, *Modern British Philosophy*, By Bryan Magee, Secker & Warburg, London, 1971.) In 1959 Russell published his book, *My Philosophical Development*, in which he said he eventually had to reject Wittgenstein because he was talking 'logical mysticism' which was anathema to his basic nature. Of course he was right. Correctly defined, logical mysticism includes religion, mysticism and unscientific gibberish, all dressed up to look like valid logical reasoning with a variety of linguistic trickery. Many aspects of philosophy in Oxford and Cambridge (and elsewhere) remain stuck in this kind of mud ever since.

with A. N Whitehead to the effect that the world of sense is a construction, not an inference. Yet even this can be traced to relativity, since Einstein made man the observer part of the observed, meaning that man contributes something to the nature of physical reality---i.e. to help with the construction of that reality---and the book was published long after both special and general relativity. It is a moot point. For the Einstein theory was the 3+1 system. The three facets of phenomena are natural; the time is, in Einstein's system, one's own local time. It means one would have to invent his time as a union between the sense of duration and external cycles before having an "objective time for general use in one inertial frame" to add to the three natural dimensions of phenomena, to complete the construction of physical reality---or the physical reality relevant to one's frame of reference.

REPLY TO SOME CRITICS ON THE WEB[20]

In the interest of learning, I am sending you the Appendices of my recent Monograph in which you will find all the answers you seek about my work. However, please understand that I am presenting a rounded philosophical theory about time, how it passes, and how it seems continuous in an attempt to solve the problem of perpetual time without the involvement of God---all of it based on Einstein's notion of time so as to link philosophy to physics by means of time alone. For a very long time I have felt that it has become possible to do so, either by time or by means of the quantum; particularly the quantum because it is the same thing as the light by which we see things as the beginning of human knowledge of the external world.

To be completely rational, epistemology can never ignore the quantum (which is also matter or small pieces of matter), as the very light by which we observe other bundles of quanta as bulky matter---a very intriguing phenomenon, or quandary, in nature. It cannot be ignored in any theory of physical reality, however conceived. To link it to philosophy is to abolish the philosophy that regards physics as 'just another way of looking at the world', and see it instead as the only way, rationally. Even the Platonic theory of Ideas becomes redundant because outside the quanta images cannot

---

[19] I am afraid, this Appendix is longer than I would have preferred, simply because my critics have to be treated with respect and answered in some detail. Discussions of ultimate reality cannot (and must not) be treated lightly.

[20] This is the corrected version of the piece posted on the Internet (in a hurry) as my reply to some critics.

exist; and the quanta are seen plainly as light---so we see how images are constructed, and by what means. The demise of Idealism is finished off.

Also, because the quantum is time-dependent (as 'energy-second'), I have made one or two comments about it. I don't think it can be the same throughout the universe because the time by which it is known on earth is peculiar to the earth---that is, provided all the universe is not subject to the 4-D geometry of Hermann Minkowski, and therefore a second here is not the same as a second everywhere else in the cosmos, according to the original Einstein theory of time. Thus refuting Minkowski is crucial. As energy-second, the quantum's energy is natural---the time is not. It is our peculiar second; and I have discussed how we make our seconds on this planet at length in the book, suggesting that it could have serious implications for the Theory of Everything.

The nature of time may have a bearing on the Theory of Everything due to the following observation about time. First of all, since Einstein was not a 'professional' philosopher he did not attempt to give the logical grounds why every body in the universe has to have its own time. [21] Unlike mathematics logic is mercilessly dry, acute and uncompromisingly factual; everything must

---

[21] Note that Einstein qualified for the noble title of 'philosopher' for a number of reasons. For instance, the difficulties over the eather or the propagation of light arose simply because the nature of physical reality had been misconceived by both physicists and philosophers---this went all the way to the quantum theory. Only a great philosopher could solve such problems and make the solution part of mathematical physics, and not as an unproven supposition, or suggestion. I will give him the title of 'The greatest Thinker', not the greatest scientists. That honour belongs to Charles Darwin, no matter what religious bigots may say against him.

be clearly defined; all conditions and methods clearly spelt out. The Einstein theory of frames is used to justify the claim that time is limited to a frame, but the technical grounds why this is so have never been made clear. That is to say, the conditions in nature that make time limited to a frame have never been clarified. Einstein could not be blamed because he was not writing philosophy.

Let me state the 'necessary logical grounds' why time is limited to a frame in a clear language (without mathematics) for the benefit of the reader or readers: time must be quantified to be useful in science and logic---let us call it 'usable time'. It is meaningless to just mention time as such. Culturally we can only use quantified time, otherwise how could we mechanise it in a clock? Now, to quantify time we have got to employ external repetitive cycles (or regular motions) in association with the internal sense of time, which is the sense of the duration of anything whatsoever, to get our usable periodicities---or time in units; so that the duration is converted to time units, or usable time; until then, time (as duration in the mind) is not usable, and can be sensed only in the passage of existence, motion or silent ageing. These units of time then become unique and applicable only to the body concerned, the body whose cycles yielded them; it is from its cycles that the units were established and so they could not be appropriate to any other body. This is not nit-picking in logic, because on this interpretation of time, all the work of cosmologists in the supposed metric of general relativity is vitiated because they use earth time; yet without this interpretation how we quantify time cannot be explained. I believe that attempts to conceive a Theory of Everything has also suffered from the use of earth time everywhere, when it is obvious that it cannot be applicable everywhere: we want to link the quantum to gravity. Yet the quantum cannot be 'a universal unit of energy' because it is time-dependent as energy-second. This must be taken into account, but it is not. Scientists just use the word 'time' and forget about its quantification and unique periodicities.

To get a fair idea of my supposition you will have to read my books about my theory, of which there are more than one. Failing that, these Appendices to my latest work will give you an idea of my philosophy. Please note that those aspects of the Minkowski mathematics you cited have no logical merit. I am questioning his basic premise. I insist that, for time alone, the i in his ict equation is not tenable; therefore his 's=ct...' deductions based on the ict equation are wrong. [22] There is no such thing as 'imaginary time'. Mathematicians often forget that mathematical symbols must have causative meanings; but philosophers never forget that. It happens to be one of the obvious differences between philosophy and science. Statistical mechanics in science (as opposed to direct one-to-one causality) overcomes the quirks and deficiencies in the behaviour of phenomena due to the absence of direct one-to-one

---

[22] At all times it should be realised that, despite the condemnation of some scientists, philosophy is important; believe me, it is very important. Einstein said he was influenced 'very greatly...' by David Hume and E. Mach. To get at what can be considered as the ultimate truth (or the truth for short, if you like), philosophers, unlike mathematicians, have to go to the roots (the logical foundations) of equations; merely repeating the mathematical symbols as written is regarded as shallow, at this level, even an insult. Let me quote part of what Russell wrote about the Minkowski formula---and you cannot say Russell did not understand the mathematics of Minkowski: "...the philosopher cannot but feel dissatisfaction with the apparently arbitrary assumption about intervals..." And again, "...there is great difficulty in suggesting any non-technical meaning for interval; yet such a meaning ought to exist, if interval is as fundamental as it appears to be in the theory of relativity..." (Bertrand Russell, *The Analysis of Matter*, Ch. xxxviii.)

causality in the nuclear and sub-atomic matter; but that does not mean the old philosophical causality can be dispense with altogether; for causality still occurs, only statistically. But logically statistical mechanics is also caused. It may not be as direct as throwing a rock to shatter a glass window; it is more like your rock going through intermediaries before reaching the glass window, so that crooked lawyers can disclaim liability; but in logic you're liable for indirectly causing the damage. This is a brief account of the type of causality now envisaged under statistical mechanics. The many mysterious behaviours of sub-atomic and nuclear matter are not without cause; for they occur because those particles exist; if they did not exist, the events associated with them through indirect causality (or statistical mechanics) would not occur. There is so much in physics crying for research in dept which scientists have neglected by relying on the fictitious Minkowski formula. Take the quantum for instance. (As energy-second, the quantum it is time dependent; it materialises periodically in accordance with out time, the units of our time as explained in the section of quantified time. If this time is peculiar to the earth, as I think, then the quantum mathematics cannot be universally applicable; and so the fear that the cosmos contradicts the law of direct causality might be misplaced. [23] I couldn't put it stronger than that. It is sufficient to indicate by this idea that more research is needed, as I suspect that the energy-second which is applicable on this planet might not be universally applicable, and so we cannot rely on the nature of "our quantum" alone to argue that direct causality is cosmically abolished---it may be so from our point of view only, for after all, how important are we. There are stars so immense that millions of our sun will

---

[23] Einstein may turn out to be right after all about this matter.

find room in them. Then we must thin of the size of our planet as compared to the sun---and the size of a human being in all that. I suspect that the quantum is not the end-piece of matter or energy in the universe at large, as opposed to what happens on this minuscule dot of a home for man.)

Back to Minkowski, he cannot hide behind the obvious lack of direct one-to-one causality to try and alter physical reality with his counterintuitive mathematics. Knowing time as it is, where is the imaginary time coming from, and what is it supposed to be like? In other words, what is the meaning of 'imaginary time'? Time, once you think of it, ceases to be any other thing than time in the clock, or quantified time as I have defined it.

The concept of imaginary time (if the reader is not aware) was invented by Hermann Minkowski; that is his ict mathematics purporting to equate space to time with counterintuitive mathematics; the i was meant to invoke imaginary time. The idea is arbitrary and therefore logically untenable. Its 'hook' which mathematicians, more theological than physicists, have swallowed 'whole' is completely unacceptable.

Warped Space And Curved Space-Time

This matter deserves a sub-section to itself because all scientists seem to have fallen hopelessly in love with it, quite wrongly, I think. Of course, we all know that the Einstein notion of gravity as caused by warped space has been proved. Then Minkowski came along to claim (merely claim) that space and time constitute one entity, and therefore when space warps, time is also warped. That idea is arbitrary and false because his ict equation upon which it is based is logically flawed. It is simply not true that time is intertwined with space and can warped so much so that you could (using the appropriate mathematics) meet your grand parents even before they were born. That is pure mythology, and comes from the Pythagorean 'Trans-migration of souls' long since discredited. If it were true none of us who are not millionaires would still be living on this planet; for wherever our grand parents might be we would join them; anywhere is bound to be better than this world!

More seriously, you will find that I know the Minkowski mathematics pretty well to even incorporate it in my corny jokes. At this level every writer is a mathematician of sorts. I even agree that he makes relativity easy to understand from the point of view of mathematics—i.e. by dispensing with the 3+1 formula and still have time inherently in space as a separate co-ordinate. But please (always) remember that Professor Eddington stated that, although useful, we must never forget that the Minkowski theory is fictitious. To me even that is unacceptable. It is not strong enough for me as a condemnation of the Minkowski proposal; useful or not, what is fictitious has no place in physics at all.

Let me digress with a brief mention of something that I know is worrying mathematicians. In discussing time rationally in that peculiar sense of physical reality championed by Ernst Mach (rather than as 'philosophy' or 'mathematics') [24], I have nothing against mathematics. In defence of Ernst Mach, let me say this: it is conceded that there are several aspects of physical reality (or science in general) better described (or written) with mathematics. Some things cannot be understood at all without mathematics. For instance, without mathematics we could not have time as we know it, because we could not state 'how much time' in numerical units; and without that the clock would not exist; civilization would be primitive. This could be a subject in Sci-fi novels---a people without clocks, and therefore condemned to live too close to nature. However, mathematics should not be allowed to dominate the entire field, for the simple reason that you could not demonstrate or write what we know about physical reality by mathematics alone; even if you could do that nobody would understand you.

The suggestion I am making is already graphically illustrated by the life story of the British Nobel Prize winner, P.A.M Dirac, the man who averred that Albert Einstein was the greatest scientist of all time because, "Only scientists like Niels Bohr and Max Planck were qualified to wipe his boots. His theories came out of the blue. They did not follow from what had gone before [and, I would add, yet they work.] [25] Dirac himself was the greatest British physicist since Isaac Newton.

---

[24] Mathematics is necessary for creating time in units (the year, for instance, as resulting from a point to a point; and there our time ends, unless we orbit the sun again); but time can never be geometricized because it is not entirely physical. The physical aspects are used merely to quantify it; but there is the inner sense of time, as the sense of duration---and how do you geometrize that? Feeling the sense of duration is as important as the hand of the clock. If the second hand of your clock moved ten paces at a time, you would sense at once that it was not giving you credible duration of time. We take the sense of duration and sub-divide it with external cycles to get time in usable units. That is the end of the relevance of mathematics in the study of time.

[25] I am quoting from memory.

However, he was regarded by his peers as a poor communicator, and sometimes incomprehensible. "...as a thinker he was unintelligible except to mathematicians. Even his fellow physicists complained that he worked in a deliberately mystifying private language..." His reply was that, "The quantum world could not be expressed in words or imagined." (Taken from John Carey's review of THE STRANGEST MAN: The hidden Life of Paul Dirac, by Graham Farmelo. Published by Faber, 2008.) Here we have both sides of the argument sufficiently elucidated. The scientific genius wanted to communicate mainly by mathematics. His equally brilliant peers objected that sometimes even they could not make out his meaning. And his response was interesting. He claimed that he was not to blame because the quantum theory was necessarily abstruse--- and we know he was right. Einstein being 'Einstein', the special theory of relativity was also difficult; and general relativity almost impossible to imagine. So both the genius and his critics are obviously quite right. Quantum theory is abstruse, no doubt about it. According to Niels Bohr "Whoever is not shocked by the quantum theory has not understood it." On the other hand, the genius has a duty to make himself understood, otherwise why bother to communicate at all? In parts of his ABC of Relativity, Bertrand Russell warns the reader not to try to visualise what he was describing in general relativity. That is one way of solving the impasse.

This means that what you state with equations (as Bertrand Russell always did) must be rendered in words too, however imperfectly. If you cannot do that your theory will never be able to stand logical scrutiny due to the absence of clarity in definitions. Here is the example I am most fond of: before Minkowski space and time were separate entities. In special relativity "they play distinctive roles", yet it works. How did they come to be one entity after Minkowski? To dwell on his so-called 'counterintuitive mathematics' raises two questions: (a)

His mathematics must be faulty, for obviously special relativity works pretty well. (b) Mathematics alone cannot demonstrate the nature of physical reality; thus the Minkowski formula does not accord with the physical reality revealed by special relativity. Hence Mach was right.

I set the Lorentz Transformation of neighbouring co-ordinates (formula) aside so as to avoid the impression that the Minkowski ict equation is based on the Lorentz Transformation and therefore acceptable. The common mistake of mathematicians has been to use the Lorentz Transformation of co-ordinates as the basis for deriving the Minkowski formula for equating space to time. This led even the great Einstein to say, "...for to every event there are as many 'neighbouring' events (realised or at least thinkable)"! [26] For Goodness sake, equating space to time will never be logically feasible because the time cannot be had without using space, otherwise how do you get it in units? Without space how do you get the year, for instance? [27] Thus the greatest logical mind in science was completely misled by the "Neighbourhood Co-ordinate" formula. Yet there is absolutely no logical method (or route) by which neighbourhood co-ordinates can lead to a natural union between space and time when the time is derived from space in the first place---unless one relies on the Minkowski mythology by supposing that, as Einstein put it, 'it is thinkable'! I do not

---

[26] From RELATIVITY, op. cit. Part I, Sect. 17. I don't know what Einstein was thinking of when he wrote this. It is an absolute bloomer! However he based it on the transformation of co-ordinates, and so he felt it was right, mathematically.

[27] How do you get any unit of time without using space? And once you get your unit of time, using space, how do you put them back together to constitute one entity---with the use of incomprehensible, abstract mathematics? This is a case where the Mach doctrine against the excessive use of abstract mathematics in the study of physical reality is particularly relevant; we have always to remember how greatly that doctrine helped Einstein.

suppose for one moment that scientists and mathematicians are unaware of the anomaly; surely everybody can see that 'thinkability' is not part of objective reality. I rather think they don't know what to do.

Mathematicians thought the Minkowski formula was a blessing as it makes things easy for them by incorporating time into space; in fact, it has turned out to be a curse, and a very serious one. The 4-D geometry is the ideal solution. Or, rather I should say, it would be the ideal solution if it were true of the natural world. Since it is not true and therefore is untenable in physical reality, it belongs to the realm of fantasy---'Dream Physics', I call it. So they don't know what to do, because they have already incorporated it into physical theory; they can do that, as I have said before, because in theoretical physics no immediate consequences flow from theories [28]---therefore nobody gets hurt when theories go wrong. Now mathematicians must swallow their pride and agree, as Kurt Gödel argued, that whatever we do certain aspects of mathematics can never be completely objective. [29] This is a clever notion; and it accords with the Platonic simile of the cave. Thus,

---

[28] No such thing as touching a button to produce results.

[29] Incidentally, I wish to point out that the Gödel formula, known as the Gödel Universe or Gödel Universes, is vitiated by the Minkowski theory upon which it is based. Every supposition influenced by the Minkowski fiction is bound to be logically flawed. It seems to me a sheer waste of intellectual effort since Eddington, Russell and Professor Whitehead told us that the Minkowski formula is logically untenable. Perhaps scholars have been encouraged to rely on Minkowski because it makes things easy for them, and also because even Einstein accepted it. Let me say that Einstein was coerced. Secondly, he knew that it could not affect relativity because whether time is the same as space or not the second is always the same; and what he wanted *is* that time is incorporated into the study of phenomena as a distinct co-ordinate.

in relativity, we must revert to the 3+1 formula used by Einstein in special relativity---if it worked there, it would work anywhere else due to the 'Two postulates'. 'Anywhere else' means in any inertial frame subject to the 'Two Postulates' of Albert Einstein. For after all we need physics (or theories of physical activity) to be effective only in an inertial frame, which is the field of special relativity applications, and where we know that the 3+1 formula works absolutely perfectly.

On the other hand, the 4-D geometry is merely 'assumed' to work in general relativity without proof. I actually believe that most of the post-relativity work in general relativity and cosmology has been falsified by the reliance on the Minkowski formula---but mathematicians only have themselves to blame, because Bertrand Russell and Professor Eddington said plainly that the Minkowski theory is fictitious and arbitrary, which meant that, logically, it was only a matter of time before it would be rejected by thinkers.

Yet they did not listen. Objections were regarded as evidence of one's ignorance of counterintuitive mathematics. All the time the proper definition of time was not even attempted. For example, where will the next second come from if the earth stops orbiting the sun? From the obvious answer that it cannot happen (or at least not just yet!), because the earth is gravitationally programmed to always go round the sun, we get the evidence that (a) the time units we use to tell 'how much time' without which our civilization could not survive, are derived from the repetitive motions of the earth; and, as such, are clearly human in origin. That is the proof that we use external cycles in union with the sense of duration to quantify time for cultural use. (b) It also goes to show that the continuity of time is obtained from the succession of time units---the repetition of the year, for instance.

The year is what we sub-divide to get all other units of time on earth. Thus no matter what mathematics are used, it is not possible to equate time (derived from space) to space again! That is contradiction in terms. The whole of post-relativity physics, the real nature of physical reality, and even relativity itself are all distorted by this mistake by mathematicians.

So I rather accept the contrary position taken by the Mathematical Society of Japan, from whose Encyclopedic Dictionary of Mathematics [30] I quote the following consensus: "Historically, the transformation formula [the equation is stated, but unnecessary here] [31] was first obtained by H.A. Lorentz, under the assumption of contraction of a rod in the direction of its movement in order to overcome the difficulties of the ether hypothesis, but his theoretical grounds were not satisfactory. On the other hand, Einstein started with the following two postulates: (i) Special principle of relativity: A physical law should be expressed in the same form in all inertial systems namely, in all coordinate systems that move relative to each other with uniform velocity. (ii) Principle of invariance of the speed of light: The speed of light in a vacuum is the same in all inertial systems and in all directions, irrespective of the motion of the light source. From these assumptions Einstein derived [the Lorentz Transformation] as the transformation formula between inertial systems $x = (ct, x, y, z)$ and $x1 = (ct1, x1, y1, z1)$ that move relative to each other with

---

[30] Published by the Mathematical Society of Japan, The MIT Press, Cambridge, Massachusetts and London, England. Ed. Kiyosi Ito. Vol. II, p. 359 B.

[31] The Lorentz Transformation and The General Transformation of Co-ordinates are discussed by Professor Eddington in *The mathematical Theory of Relativity*-- Sections 5 & 15. In any case, the Japanese mathematicians did not think much of the Lorentz Transformation, neither did I, and it is not strictly relevant here either.

uniform velocity v along the common x-axis. This was the first step in special relativity, and along this line of thought, Einstein solved successively the problems of the Lorentz-Fitzgerald contraction, the dilation of time as a measure of moving clocks, the aberration of light, the Doppler effect, and Fresnel's dragging coefficient." The time Einstein used, according to him, was the Lorentz local time, provided, it can be defined as 'time, pure and simple'---meaning it is all there is of time; this qualification is very important. (Apparently it can be so defined, for it did not hinder his work until Minkowski intervened.) It means every time is somebody's local time. To overcome that we have learnt to mechanise our time in the clock for general use---but, and this is the crucial point, it is based on the earth's motions. So earth time as derived from the earth's regular motions is all the time we have or can have. This is definitely the view from Einstein; and that was the end of the matter rationally, as Einstein would have it.

Yet, by using concepts like the 'homogeneous Lorentz group', 'time reversal' 'space reflection', 'parity transformation', 'the proper Lorentz group', and so forth, none of which made the original Lorentz formula satisfactory, as the Japanese mathematicians aver, cosmologists, the interpreters of general relativity and pure mathematicians are propounding impossible theories about time and call them Einstenian, but mostly inspired by the Minkowski theory that Russell and Eddington described as arbitrary and fictitious. I hope I make myself clear to avoid any misunderstanding.

By the Minkowski theory time travel is said to be 'a scientific possibility'. I am afraid that is not true. All notions of time travel are sheer humbug, because Minkowski was wrong. I have to add that it is quite unacceptable to try to conceal logical errors in thought with mathematics as Minkowski has done. Einstein is not to blame; he was literally coerced by mathematicians to accept the Minkowski formula, but either way relativity is not affected.

There is credible evidence to justify the claim that relativity is not affected whether time is regarded as the same as space, or independent of space. Even I should say that the evidence is not only credible but strictly logical. In the special theory of relativity Einstein made time independent of space. Why didn't he go back to make the time the same things as space in the 4-D geometry after he adopted the Minkowski formula? To me, it is because good old Einstein was no fool. He despised the 'superfluous learnedness' of the mathematicians who were tampering with his theory; yet he needed the support of the scientific public, the majority of whom were coercing him to adopt the Minkowski theory. At first relativity was universally ignored; so he acquiesced; but he was no fool. I believe he knew that either way relativity is not affected. Thus he left special relativity as he originally conceived it---where space is separate from time.

But whether time is the same as space or separate from space, the second is always the same---time is always the same; and Einstein wouldn't miss that point even without his brains. The difference between the two versions of time is philosophical, and it is this: thanks to Einstein we now know that every time is somebody's time; that there is not one (overriding) system of time that covers the whole universe. A second here is not like any other second anywhere else. The Lorentz concept of local time was called 'time, pure and simple' by Einstein, meaning that it is all the time there is, or can be had.

Considering how fundamental time is in human affairs, this is a philosophical concept of a revolutionary kind--- there are as many times as there are bodies. Russell put it most succinctly (and I quote from memory): "...there is no longer a universal time which can be applied to any part of the cosmos without ambiguity..." It means our time cannot be applied anywhere else. He's right. So far we know only of the time we have created as our local time to suit the earth's motions. Yet time goes to the roots of our existence. It is closely associated with 'Being' or 'Existence'. There is a natural aspect of time in our conceptions. Professor Eddington called it 'the internal time-sense', being the sense of duration. But we also know that it requires points; therefore it cannot be the same as 'Being'. For it could not have been invented without using points; sentience was required.

Thus we must look for an aspect of life that can be mathematically linked to repetitive external cycles (the years, for example) to yield units of time that accords with physical reality (the most obvious is Day and Night), and can also be mechanised in a clock, which, once achieved, should be seen as man's greatest intellectual (scientific, mathematical and philosophical) creation or invention---and Einstein led the way to the most logical theory ever; that is the reason I maintain that time was Einstein's greatest achievement, for that is how we can logically link physics to philosophy by means of time.

Hence in post relativity physics time must be based on our roots, and I think it is now seen as such. But how? The mechanics must be explained. Well, time is now logically conceived as something whose roots in our minds are based on the sense of duration (or the capacity to experience duration as an internal sense of time, as Professor Eddington put it: that is, of the impressions, images, and so forth, of things enduring, since it takes time to endure or linger), and therefore part of the physical or physiological mechanism for memory---if memory is defined as 'the capacity to repeat'. [32] And

---

[32] This goes to the roots of our existence because it is part of the mechanism by which we gain knowledge and remember it---which is the sum of the contents of the human mind; part of that knowledge

that is how we link physics to philosophy. The Minkowski formula contradicts this supposition; as such, his theory would be the greatest achievement of the human mind (linking physics to time via space); but then he based his theory on imaginary time co-ordinates, and therefore it is regarded as fictitious and arbitrary (by Professor Eddington and Bertrand Russell, no less).

I confess it is (or was) technically difficult but I have somehow managed to show why the Minkowski formula seems to work. This is just an elaboration of the gist of the above paragraph. But for your (and everybody's) benefit it will do no harm to repeat it. The question is whether time is secular and originates on this planet, or it is generally in the universe to be invoked with the appropriate mathematics (and mathematicians seem to prefer the cosmic interpretation that implies the existence of God); secondly, whether i can invoke such a time. It cannot because imaginary time does not exist anywhere except in a dream. However, time (the second) is always the same in our minds whether the time is seen as separate from space or part of 4-D geometry; and that is what Minkowski exploited. The second is always the same in the mind, but how does it get there? Again, I have shown the technical methods in my works as a link between the sense of duration and external cycles---the years, for example---and Professor Eddington said much the same thing (ibid, Ch. 1.8.) Professor Eddington wrote: "The rough measure of duration made by the internal time-sense are of little use for scientific purposes, and physics is accustomed to base time-reckoning on more precise external mechanisms." (ibid, Ch. 1.8---p23, my italics.)

---

by which we live is the concept of time, of things lingering. It takes time to linger; added to repetitive external cycles, we get time units to mechanise into clocks.

Here is a brief explanation of the idea that Duration x external cycles = time in units, for usable time is in units only, known as 'quantified time': the year is just one unit of time, and all other known units of time are derived from the year with points, thus making them also discrete, including the cesium units, since they have to be related to the second to make sense. To begin with, let us assume there are no clocks: now suppose you see an image on TV, then it goes away after a while. How long it was there is its "duration" in sense or the mind, otherwise known as 'the internal sense of time', or 'the internal time-sense', as Professor Eddington put it. It must be clearly understood that duration implies the passage of time---i.e. during the period (or the life) of an event. But obviously it is not enough; it is not the time you can mechanised in a clock for general application, again as Eddington put it. For cultural purposes [33], something else must be added to the duration, namely, it must be converted to units of time. This is easy to do, for we know how we get the earth-year as a unit of time---round the sun as a cycle. In fact, the year is our basic unit of time, metaphysically. To have more units (or years) we go round the sun again, otherwise there are no naturally occurring time units, or years. Strictly speaking, this is not a theory. The process of creating time units to mechanise into clocks is the metaphysical origin of time as a union between duration and how it is broken down into units---or cycles.

Thus, to convert duration to time units, you will have to use repetitive external cycles like the earth-year. [34] This

---

[33] For any purpose where time is to be cited (as the additional co-ordinate of relativity, for instance), you need to have the time in numerical units, which can only be achieved with external cycles.

[34] This is easy to understand. You can even tap your finger, and say, for instance, the duration (or the life of the relevant event) was for so many taps of the finger.

procedure will be the same for any sentient beings anywhere else in the universe; we can only use repetitive cycles to create time in units. That is the only logical way to obtain time in quantified units---otherwise time is 'silent ageing', 'silent motion', or 'the silent passage of existence', all of which are useless to science and logical thought. Without mathematics logical thought (in abstraction or in any depth) is not feasible [35]; and you need to apply mathematics to duration to get time in numerical units suitable for logical thought. In sleep or coma time will be passing by. But when you come to and want to know the time or how long you've been senseless, you will need mathematics based on some kind of repetitive motions or cycles to be able to have the time in numerical units.

We on earth use the earth-year as sub-divided down to the seconds, or the cesium units, to determine the time "during which" the image was there. The term 'during which' means duration, but it is not enough as time. [36] You will have to relate it to some of the sub-units of the earth-year to get the appropriate time. (You will say the event was 'so-and-so long'; that so-and-so length is obtained elsewhere and applied to the event. Metaphysically there is no other way. [37]) Thus we apply some of the earth's sub-cycles to 'duration' to get the

---

[35] We hear of the invention of points being necessary for mathematics. In fact, basically, it was a logical invention necessary for mathematics, since mathematics is logical thought in abstraction---e.g. for handling massive volumes and representational reasoning where you cannot see what you are talking about.

[36] Because images, impressions, events and so forth, can linger in the mind, they are obviously connected to the mechanism for memory, which is defined in science as "the capacity to repeat".

[37] That is why 'time' and 'the application of time' are two distinct operations of the mind, but often they are conflated leading to unnecessary mysteries about time, as discussed earlier.

time for it---to get the time 'during which' it was there; and that means converting duration to time by linking it to external cycles. The external cycles themselves do not constitute time, either. They are given durations, or periods of mental lengths (during which they were there), before they can constitute time: a month is longer than a week; and so forth. A second is shorter than an hour. It is by the sense of duration (during the life of an event, an image or an impression) that you can determine which unit of the external cycle to apply to it to get the time, or the number of cycles it was there. The two statements (time and number of cycles) are exactly equivalent. [38] So that you can say it was there for one cycle=one year. Or apply any of the sub-units of the year's cycle to it. Due to our use of clocks we have forgotten that this is how we created and mechanised time for the clock/Calendar system. The metaphysical question is whether there is any other time. Well, the passage of existence is regarded as time, but it's not quantified or usable time. My own opinion is that the method described is the only logical system that any sentient beings in the universe will use to get their quantified time because logical thought must definitely be universally the same everywhere as we have it on earth---for it is a process of reasoning about percepts. Nobody can live rationally in any part of the universe unless he or she adopts reasoning according to percepts, and that is what we call logical thought. It may get more complex and mathematical with increasing volumes (and in non-demonstrative inferences), but ultimately it must be based on percepts. Even Idealism, before it was successfully refuted, was somehow related to percepts: if you cannot see anything the question as to whether it is mental or physical will not arise---simply because the 'it' will not be

---

[38] For example, the numbers of the earth's cycles or orbits round the sun are known as years, or periods of time.

known. If the Irish philosopher, Bishop Berkeley, ever saw anything (say, the pen he wrote with, the paper he wrote on, the table and chair in his study), then there is no argument. In any case, Berkeley rather confirmed the existence of the quanta hundreds of years ahead, without knowing it, as Bertrand Russell has pointed out in his History of Western Philosophy. The most interesting refutation of Idealism, of course, is the quip that a train at a station that can be seen to have wheels cannot be said to lose its wheels when in motion just because the passengers are not seeing them, or looking at them as the Idealist ideology requires.

About myself and reactions to your criticisms, I confess I felt a little sad, not really annoyed but rather sad, that you should mention the Minkowski mathematics to me. At this level it is most unfair to assume that I could be ignorant of the Minkowski mathematics. The truth is that mathematicians have overlooked the caveat of Professor Eddington (in his monumental opus, The Mathematical Theory of Relativity), to insist that Minkowski has changed the nature of time with his formula. To repeat: he said plainly that the Minkowski formula would be ideal for describing phenomena, but, while mathematicians may consider it as useful, they must not forget that it is arbitrary and fictitious (Mathematical Theory of Relativity, Ch.1.1.)

Yet what is the situation now? We find that what Professor Eddington and Bertrand Russell have both described as arbitrary and fictitious is making mathematicians shameful because they allow their works to be guided by it, and out of which such mythologies as time travel become scientifically possible. As always with poor old mankind, people show how clever they are as books about such subjects sell millions and my books are ignored! Even the journals will not publish my papers, and judging from what you say, and seem to believe in so completely, I am not surprised. If Minkowski was right why didn't Einstein go back to amend special relativity? [39] He was coerced to use the 4-D geometry in general relativity to make it easier to understand---but he was wrong. [40] As a result of which general relativity is now in a hopeless mess. Yet again, either way the basic postulate of general relativity (the curvature of space for gravity all the way to inferences about black holes) is not affected. Einstein was no fool! Footnotes of his theories will come to replace the footnotes of Plato's theories in philosophy. That is my prophecy. Einstein must have known that, intellectually, he's something like God.

Finally, I have already indicated that we can link physics to philosophy through the relativity concept of time; we can do the same thing, again, through the same Einstein's concept of the quantum, or his "Light Quanta" theory, too. Originally he called it 'a hypothesis'. But it is no longer a hypothesis, not even a theory, as scientists have confirmed it as a fact of nature through QED. So the man I call the only God we know did a lot to justify his divine appellation a hundred times over.

---

[39] Another question is when cosmologists and mathematicians are going to realise that earth time is not applicable to any other world, frame or metric, outside this planet. All the work they have been doing in general relativity since Einstein is vitiated because Minkowski is wrong and our time is not applicable to the metric of general relativity---it is a different frame, or world, 'pure and simple' as Einstein would put it.

[40] What all those lazy (and also probably religious) scientists want is that time does not have to be added to phenomena in the 3+1 formula as Einstein proposed; they rather prefer time as inherently part of space naturally. But that is based on the Minkowski formula, and the Minkowski formula is logically invalid. It seems we are speaking about two different worlds: one which obeys the laws of logic, and another that lives in the imagination of some scientists because of the Minkowski false theory. Yet since the deaths of Einstein, Eddington and Bertrand Russell, they have incorporated the latter into theoretical physics---it is wrong, but who is to check them?

The quantum is seen as light. A single one will be a solitary speck of light of a specific hue; and as small as they are, scientists have invented a machine for counting them one by one (QED is said to be the most well-established of scientific theories, and its all concerned with quantum interaction). En masse we call them quanta---or the smallest pieces of matter that can exist. We can link physics to philosophy through the quantum because it is light, and we have always thought we see only by means of light. That thought is rather a misconception. In actual fact, we never see things at all; we see only their images. [41] Physiologically it is quite impossible to perceive anything. Seeing occurs in the brain not on the eye. How, for instance, can we fit a house physically into the brain's tissues? Rather we think we see a thing because its lights consisting of these very small quanta reach the eye with its exact image, and thence to the visual cortex in the brain. So we see only the surface lights of things, not the things themselves. Another philosopher conceived his theory of vision just to account for this very fact. It is precisely what the Irish philosopher, Bishop George Berkeley, stated in his philosophy, although he interpreted it wrongly, as Bertrand Russell put it: "Berkeley advances valid arguments in favour of a certain important conclusion, though not quite in favour of the conclusion that he thinks he is proving. He thinks he is proving that all reality is mental; what he is proving is that we perceive qualities, not things, and that qualities are relative to the percipient." This is a permanent proof that we do not see things; we see only their surface qualities in the form of their exact images as constituted by their light emissions, or radiations, by means of quanta---the smallest bits of matter in existence.

---

[41] Plato conceived his Theory of Ideas to account for this fact.

We interpret this as meaning that we see only the smallest parts of matter that, by their nature, can never stand still and always radiating about through their interactions with the electrons of matter. So we see matter, and yet do not see matter. We see only matter's smallest parts flying about---but they are also matter. [42] The philosophical importance is that pieces of matter do actually fly off things with their exact images; when we capture some of them on the eye we see them, with all their colours because the quanta are naturally coloured. There are intricate technicalities in all this; physicists will not put it so bluntly or crudely; they will have to add a number of fine qualifications. But to philosophers, that is enough to work with. Capturing quanta from things to see them means, as Bishop Berkeley supposed, seeing the surface qualities of things and not the things themselves---the main reason is that all seeing is 'Tele-Vision'. When you are very close to a thing it is difficult to see it properly; if it is too close to the eye you might not be able to see it at all without confusing blurs. Seeing by means of quanta is precisely like photography. The only problem is size.

Size presents a special problem that can only be settled with a dose of speculation as opposed to solid facts. But the speculation is based on the facts given above, and they have been proved more than sufficiently. All that is required is that the inferences based on them should be logically valid.

---

[42] Because of QED we cannot (at least, I cannot) escape the fact that the quanta are the smallest pieces of matter out of which all other forms of matter have emerged through coalescence. The Scientific Industrial Complex has found another way of spending our taxes, so scientists can go on (who can stop them?) building larger and larger 'Atom Colliders' to split atoms in search of the basic building blocks of matter; but for me, my notions of the origin of all forms of matter begin and end with the quanta, the particles of light, thus making light 'What Is' in logic.

Now, because it involves the quanta, seeing is obviously like the electronic scanning of (or in) computers. [43] We imagine that human vision occurs atom by atom, since the quanta are the smallest sub-atomic particles---they interact with matter through their sub-atomic parts, particularly through the electrons of atoms of matter. This well-known electronic process gives a clue as to how size is handled by the brain. The thousands of pages in a computer file are not spread physically in the computer as we spread papers on a table. Similarly, even a single A-4 paper cannot fit into the brain's 'bloody' tissues. I mean, let's face it, when surgeons cut open the brain they see only bloody tissues; yet these same bloody tissues give us awareness of non-bloody percepts; and we know that is possible only at the sub-atomic level of physical reality. At this point a certain amount of speculation creeps into our thoughts. We believe that seeing the paper (of any colour) involves electronic processing of quanta in the brain, where the bloody tissues our surgeons see are not bloody at all but part of the electronic processing of quanta in things and people's heads. The paper's size is electronically scanned end to end. Since it has its own colour, when it ends, its colour will also cease---and another thing's colour will take over. Let us suppose that the paper is white. At every end of it (all four corners of it) different things' colours will take over. We call the four white corners 'the paper'. The other colours would belong to something else, because the paper has ended, and something else must be there not void. Even void has its own colour to identify its presence. This is how one thing

---

[43] I have to sound the warning that, although the computer can give relevant examples regarding the operations of the human brain, all such examples are copied from the brain. Let me explain this contradiction in terms. Scientists take suppositions from many sources and test them; some work, others don't; but even those that work are poor imitations (and certainly less complex) of how the brain actually works.

is one thing, and another thing is also another thing in vision. Needless to say, given the nature of human motives, confusion in vision can be induced through this method of visual perception---in magic, for instance. The white sheet of paper will stand out; when it ends, its colour will also end to show that it is no longer there before the eyes. The end of an image is marked by change in colour. As far as the brain is concern, the process of seeing the white piece of paper is like the computer scanning it minutely atom by atom at the quantum level, which is very small indeed.

In this way objects of any size can be visually perceived (electronically scanned), without having the things (of so many different sizes) physically lodged in the brain tissues. Size in vision is determined by volume, shape or form, colour, space and position---all of which can be interpreted as 'different colour patterns'. [44] The light we have to shine on things to see them create billions of quantum points of emission, "scanning" [45] the objects of vision as if with a torch in the dark but from billions of

---

[44] There is scientific support for this because the best scientific definition of size in vision is called "pattern recognition"---sometimes the phrase is used to represent the actual process of vision itself.

[45] Scanning is not the correct word; but there is no correct word. What is involved is so unique there are no similarities. You have to pity those who write about the quantum with our "crude" traditional means of communication. The nearest description is to say the individual quanta are so small that the total number required to give vision of the wide open sky would fit on to the sharp point of a needle with copious room to spare. And they will persist, making it look like scanning. So long as the thing is there, the quanta radiating from it will persist; therefore vision will occur---the eye is bombarded by the quanta from the thing. Remember that they are trillions x trillions x trillions; they are persisting; they are radiating from all aspects of the thing; thus they will continue to give vision of the thing. The visual process is better describes as the absorption of quanta, the quanta from things---we do not see the images of objects such as is supposed on the old Platonic theory. We absorb their material signals direct from them; if they are big, the visual process is like scanning their surfaces---not quite, but almost like that.

points, thus conveying size in vision. So light does not illuminate objects for us to see them; rather they cause objects to emit lights that carry their own exact images---thus making the Platonic Theory of Idea redundant, as a bonus. Let me repeat this because the Platonists are pretty stubborn: the light source does not illuminate the objects we see. We need the light all right, but it does not illuminate objects. By the miracle of quantum computation and interactions, this is not at all surprising. In fact there is no such thing as illumination in the universe. According to Niels bohr Light is "Transmission of energy between material bodies at a distance". Thus we should regard the quanta from the light source as interacting with the atoms of the things we see, rather than illuminating them. Emission and absorption of radiation is corpuscular. In the absence of illumination in the universe, Plato was plainly wrong.

The inevitable conclusion is that both science and philosophy (or physics and philosophy) have identified one entity as the ultimate cause of the nature of physical reality, or "What Is", namely: we see the world through the quanta; also in all its physical analysis of the nature of the external world, physics has identified the quanta through QED as the cause of all physical reality in its multitudinous forms. So physics and philosophy are linked in the existence of the quanta, another of Einstein's discoveries. He has therefore done more than we expected of God---even Plato is no longer interesting.

On the other hand, if we are to replace the Platonic Theory of Ideas with the new concept of "Quantum Signals" issuing from things to enable us to perceive them visually, then we have to adopt the corollary, which is "Coded Signals In Perception", meaning that things are given codes in the brain for purposes of memory (and all cognitive processes) and not the things themselves physically, since all things are just too large to fit into the brain as they are seen externally.

This sounds confusing, so I will do my best to explain it. I think electronically coded signals cause vision, and that it is the reason we can dream of objects and of events with the eyes closed. REM seem to indicate selectivity and scanning as in the computer. It is interesting that scientists have found that vivid dreams are particularly associated with REM. [46] The electronic coding may have internal and external aspects. Internally, it is obviously part of the mechanism for memory. Externally, the brain must have created categories of 'perceptive images' (images of classes and groups of objects) over several centuries: the figure of a person, the flight of birds, quadruped motions of animals, shapes and forms of objects, and so forth---a long list of categories, literally infinite. Anything new will get its own category; anything known will have its established category already in place; thereafter associated objects will be added to it. What is on two feet, would be assigned to the category of two-footed creatures, including people; what is flying, to the category of flying objects, and so on. Thus we imagine that when something is apprehended, the brain is able instantly to place it in its appropriate category and recognise it. Internally, the codes of things come into play when the thing is invoked (through the appropriate stimuli); so that however big, it can be seen in the mind's eye, because the coding system is electronic on the quantum level. These are all guesses or speculation, but without them vivid dreams cannot be explained. We are trying to imagine how the brain works; nobody has any cast-iron proof.[47] But what we

---

[46] By the Sleep Research Centre at Loughborough University, for instance.

[47] However, as his final theory of the human mind, Bertrand Russell said that when a physiologist is examining the brains of a patient, what the physiologist is seeing is in his own brain---see the Chapter, My Present View of the World, in his book, My Philosophical Development. This theory of the mind implies a perceptive process that relies on signal codes in visual perception: we imagine that the

have found so far sounds credible. The computer can help, since it works through the system known as 'pattern recognition'; and, as I have said above, in the matter of deciding how size appears in vision, we assume that the quantum signals from things behave as if they are scanning the objects from end to end. So the size of any object will extend to the end of the thing's colour. In this way it is believed that size is determined by shape or form, position and colour---all of which can be abstracted to 'different colour patterns', because even 'the void' has to have its own colour---its colour has to be different from those around it.

There are aspects of the brain physiologically established from scientific medicine, computers, and human behaviours that suggest that the above theory may be close to the truth of how the brain actually functions. Nevertheless, even if the suggestion is proved scientifically, it will take centuries for people to digest these very difficult ideas based on relativity and other ideas of Einstein which are also not yet properly understood even by eminent Professors.

One of the difficulties comes from the fact that, if the quantum is a product of our time, as erg-sec, and the units of our time that give rise to the quantum, are also products of the human mind (through a link between the sense of duration and external cycles), then the nature of human life ought to be re-defined because it is not properly conceived either in science or philosophy. For the quantum which is the basis of all matter as far as man is concerned, may not be naturally in existence throughout the universe in the form it appears to us, but rather materialises through our unique way of moulding

---

patient's brain parts invoke the appropriate codes in the physiologist's brain to make him see them---as he would do in dreams with his eyes closed.

elements of nature to suit our unique nature; and therefore the 'Copenhagen Interpretation' can be adapted to mean that the strange behaviours of sub-atomic matter may be due to the strange manner in which the human mind is 'generated continually' and interacts (and interferes) with nature, namely, not constituted as a solid matter (or mass) but put together by fleeting and highly perishable impressions at the electronic level---and always growing, changing, interacting and inter-communicating through the neurons, for instance.

It seems nature does not see us as special, or as human beings, but mere neurological robots with no right to claims of superiority or any metaphysical pretensions; perhaps we are more efficient than other animals in nature---well, maybe that implies that we are somewhat superior in a way, but I doubt that that is metaphysically significant. Nobody can have the last word because nobody knows what the origin (or the purpose) of life is.

The importance of what Einstein did, even without knowing it, is this: since the dawn of civilization philosophers have been trying to interpret the world to know what it is made of, or how it is constituted. Out of their inquiries scientists arose to claim that the philosophers have got it all wrong. And they started their own lines of enquiry, the chief part of which is the physical analysis of the external world, known for short as 'physics'. Einstein did not know that the philosophers and scientists have reached a stage in their enquiries where all their theories coincide in the discovery of the quantum---as the most credible candidate of What Is. And the miracle is that man does not even have to infer this idea from any complex theoretical postulates from scientists or philosophers. We see the quantum plainly as light. Of course, it has taken a long time to come to this conclusion, more than one hundred years in fact. Einstein himself did not know it---unaware of the significance of what he had achieved. That is because

105

the work of interpreting his ideas to come to this conclusion was difficult. [48]

The most important thing was the new secular theory of time. Until then time was so mysterious that it had literally become the last hiding place of God after Charles Darwin. I can imagine religious leaders smiling smugly at the increasingly secular theories coming out of physics. But as something that originates from this planet (because there is no longer a universal time), and can be seen as a union between the sense of duration and external cycles, time is liberated from the religions; but it happens to be the most important aspect of life, second only to how the life itself came to be in the universe of inanimate matter, and without which no civilization (to sustain life) could have been possible. With this secular theory of time, added to the theory of quantum, we can now agree that man's science will die with him when the earth ceases to be habitable. Nature is far from uniform; the theories that work for us here on earth will die with us; so let's make the most of our good fortune in the work of Albert Einstein.

With the above thoughts in mind, let me hurry to point out that the importance of linking physics to philosophy is to defeat the murderous peddlers of stupid religious myths from all the religions. Since philosophers are strictly logical thinkers, it will put an end to all that buffoonery nonsense from callous religious bigots together with their dung-headed leaders, who claim that science is but one of many ways of studying the world and no superior

---

[48] For nearly half of a century, all my numerous papers to the academic journals were rejected---the editors could not understand what I was saying, and I don't blame them. Even Einstein did not know what he had achieved in this regard. What worries me is that with the Minkowski formula, his lofty theory of time is vitiated; that is why it is necessary to point out that the Minkowski theory was condemned even by Professor Eddington as 'arbitrary and fictitious'.

to any other. Once we eliminate such murderers who promote suicide bombings as passports to Heaven, the only remaining problem, as I see it, is how best to control science rationally to man's utmost benefit all over the planet, so that a scientific forum, organised on the lines of the UN, will come to exist---hopefully---to eliminate mad scientists, who will come, oh yes, they will rise, probably disguised as (religious) scientific prophets!

REFERENCES

(Many sources are given in the text. The following are among the major books and papers on the subject.)

ALBERT EINSTEIN (1879-1955)---SPACE TIME, an article in the 1926/27 (13th) edition of the Encyclopaedia Britannica. Also, RELATIVITY, in the same edition.

--------NATURE No. 106, 782, (1921), almost the whole issue was devoted to the confirmation of Einstein's new theory of gravity.

---------The Meaning of Relativity, Princeton University Press, 1966.

---------The Evolution of Physics, (With Leopold Infeld) Cambridge 1838.

---------RELATIVITY, Routledge Classics, London and New York, 2001.

HERMANN MINKOWSKI (1864-1909)----He first mentioned his supposition in a lecture in cologne, known as Raum und Zeit (Space and Time) Cologne 21st September, 1908.

---------- Herman Minkowski AdP 47, 927 (1915)

----------Herman Minkowski, Goett. Nachr., 1908 p53. Reprinted in Gesammelte Abhandlungen von Herman Minkowski. Vol. 2, p352. Teubner, Leipzig 1911.

BERTRAND RUSSELL, FRS (1872-1970)----Our Knowledge of the External World, George Allen & Unwin, 1922.

-------- Mysticism & Logic, George Allen & Unwin, 1976: a collection of important essays first published in 1917.

--------ABC OF RELATIVITY, George Allen & Unwin, 1958 (recently revised by Professor Felix Pirani---first published in 1925.

--------History of Western Philosophy, George Allen & Unwin, 1946.

--------My Philosophical Development, George Allen & Unwin, 1958.

--------The Analysis of Matter, George Allen & Unwin, 1927.

MORRIS KLINE: Mathematics in Western Culture, Allen & Unwin, London, 1954.

SIR ARTHUR STANLEY EDDINGTON, FRS (1862-1944)

-------The Expanding universe, University of Michigan Press, Ann Arbor, 1933

-------The Combination of Relativity Theory and Quantum Theory, Communication of the Dublin Institute of Advanced Studies, Dublin, 1943.

---------The Mathematical Theory of Relativity, Cambridge, second ed. 1930.

---------The Nature of the Physical World, Ann Arbor, Michigan, 1958.

--------Philosophy of Physical Science, Cambridge, 1949.

-------The Theory of Relativity and its Influence on Scientific Thought, Oxford, 1922.

-------Space, Time and Gravitation, Cambridge, 1920.

SIR JAMES JEANS, FRS: Physics and Philosophy, Cambridge, 1942.

-------The Mysterious Universe, Cambridge, 1930.

-------The New Background of Science, Cambridge, 1933.

PROFESSOR A.N. WHITEHEAD: The Concept of Nature, Ann Arbor, Michigan, 1957.

-------Science and the Modern World, Cambridge, 1922.

-------An Inquiry Concerning the Principle of Natural Knowledge, Cambridge, 1919.

-------Nature and Life, Cambridge, 1934.

-------Process and Reality: An Essay in Cosmology, Cambridge, 1929.

-------Essays in Science and Philosophy, Rider & Co., London, 1948.

-------The Principle of Relativity, Cambridge, 1922.

Professor BANESH HOFFMANN: The strange Story of the Quantum, Dover Pub. Inc. New York, 1959.

Professor STEVEN F. SAVITT (ed.) Times Arrows Today: Recent Physical and Philosophical Work on the Direction of Time, Cambridge, 1995.

CHARLES A. FRITZ: Bertrand Russell's Construction of the External World, Routledge & Kegan Paul, London, 1952.

Professor JEREMY BERNSTEIN: Albert Einstein and the Frontiers of Physics, Oxford, 1996.

Professor RICHARD FEYNMAN: Lectures---The Character of Physical law. MIT Press, 1967. There are several volumes of the Feynman lectures and they are all worthy of serious study.

Abraham Pais, "Subtle is The Lord: The Life and Science of Albert Einstein", Oxford, 1982. Professor Pais has methodically provided details of almost all the original papers relevant to relativity. His list is so exhaustive I don't know of a better one anywhere.